ISSN 0044-2267

ZEITSCHRIFT FÜR ANGEWANDTE MATHEMATIK UND MECHANIK

INGENIEURWISSENSCHAFTLICHE FORSCHUNGSARBEITEN

BAND 61 1981 HEFT 11

AUS DEM INHALT

HAUPTAUFSÄTZE

AKADEMIE-VERLAG · BERLIN

ZAMM Bd. 61 Nr. 11 S. 541—604 Berlin, November 1981

EVP 18,— M 34115

INHALT

Hauptaufsätze

Kleine Mitteilungen

Wir bitten, **Manuskriptsendungen** zweifach (Original und eine Kopie, sprachlich einwandfrei, Formeln mit Maschine oder in Druckschrift geschrieben) an folgende Anschrift zu richten:

Zeitschrift für Angewandte Mathematik und Mechanik
Institut für Mechanik der Akademie der Wissenschaften der DDR
DDR-1199 Berlin, Rudower Chaussee 5

Zu den Arbeiten, die als Hauptaufsätze bestimmt sind, ist auf gesondertem Blatt eine Zusammenfassung von 5 bis 10 Zeilen in englischer und (möglichst) deutscher und russischer Sprache beizufügen. Ausführliche Hinweise für die Autoren, um deren strikte Berücksichtigung gebeten wird, finden sich im Anschluß an das Inhaltsverzeichnis des Jahrganges 60 (1980).

Die Autoren erhalten von den Hauptaufsätzen 75, von den Kleinen Mitteilungen 25 Sonderdrucke ohne Berechnung, darüber hinaus weitere Sonderdrucke gegen Berechnung.

Bestellungen sind zu richten

- **in der DDR** an den Postzeitungsvertrieb unter Angabe der Kundennummer des Bestellenden oder an den AKADEMIE-VERLAG, DDR-1086 Berlin, Leipziger Str. 3—4
- **im sozialistischen Ausland** an eine Buchhandlung für fremdsprachige Literatur oder an den zuständigen Postzeitungsvertrieb
- **in der BRD und Berlin (West)** an eine Buchhandlung oder an die Auslieferungsstelle KUNST UND WISSEN, Erich Bieber OHG, D-7000 Stuttgart 1, Wilhelmstraße 4—6
- **in den übrigen westeuropäischen Ländern** an eine Buchhandlung oder an die Auslieferungsstelle KUNST UND WISSEN, Erich Bieber GmbH, CH-8008 Zürich/Schweiz, Dufourstraße 51
- **im übrigen Ausland** an den Internationalen Buch- und Zeitschriftenhandel; den Buchexport, Volkseigener Außenhandelsbetrieb der Deutschen Demokratischen Republik, DDR-7010 Leipzig, Postfach 160; oder an den AKADEMIE-VERLAG, DDR-1086 Berlin, Leipziger Straße 3—4

ZEITSCHRIFT FÜR ANGEWANDTE MATHEMATIK UND MECHANIK
Herausgeber und Chefredakteur: Prof. Dr. Günter Schmidt. Redaktion: Dr. Winfried Heinrich, Dr. Horst Weinert, Dipl.-Math. Friedhild Dudel, Helga Rühl, Institut für Mechanik der Akademie der Wissenschaften der DDR. Verlag: Akademie-Verlag, DDR-1086 Berlin, Leipziger Straße 3—4; Fernruf: 2236221 oder 2236229; Telex-Nr.: 114420; Bank: Staatsbank der DDR, Berlin, Kto.-Nr. 6836-26-20712. Anschrift der Redaktion: Institut für Mechanik der Akademie der Wissenschaften, DDR-1199 Berlin, Rudower Chaussee 5; Fernruf: 6802841

Veröffentlicht unter der Lizenznummer 1282 des Presseamtes beim Vorsitzenden des Ministerrates der Deutschen Demokratischen Republik. Gesamtherstellung: VEB Druckerei „Thomas Müntzer", 5820 Bad Langensalza. Erscheinungsweise: Die Zeitschrift für Angewandte Mathematik und Mechanik erscheint monatlich. Die 12 Hefte eines Jahres einschließlich Tagungsheft bilden einen Band. Bezugspreis je Band 360,— M zuzüglich Versandspesen (Preis für die DDR 216,— M). Bezugspreis je Heft 30,— M (Preis für die DDR 18,— M). Bestellnummer dieses Heftes: 1009/61/11.

AN (EDV) 35937

ZAMM 61, 541 – 545 (1981)

J. Wolska-Bochenek / L. v. Wolfersdorf

On Some Generalized Free Boundary-Value Problem for a System of One-Dimensional Diffusion Equations

Es wird ein in der Metallphysik vorkommendes freies Randwertproblem für ein System von zwei eindimensionalen Diffusionsgleichungen nach dem Vorbild von A. Friedman durch Zurückführung auf ein System zweier nichtlinearer Volterrascher Integralgleichungen behandelt.

A free boundary value problem for a system of two one-dimensional diffusion equations occurring in metal physics is dealt with by reducing to a system of two nonlinear integral equations of Volterra type following a known method due to A. Friedman.

Исследование свободной граничной задачи для системы двух уравнений диффузии в одной пространственной переменной появляющей в физике металлов как у А. Фридмана сводится на систем двух нелинейных интегральных уравнений Вольтерра.

I. Introduction

Let us consider the system of two partial differential equations

$$\sum_{j=1}^{2} d_{\alpha j} \frac{\partial^2 u_j}{\partial x^2} - \frac{\partial u_\alpha}{\partial t} = g_\alpha(x, t), \quad \alpha = 1, 2, \tag{1}$$

in one space dimension, where the constant square matrix $D = [d_{\alpha j}]_{2\times 2}$ and the functions g_1, q_2 are given. If we assume

$$d_{\alpha j} > 0, \qquad \det D > 0, \qquad \alpha, j = 1, 2, \tag{2}$$

then the eigenvalues λ_1, λ_2 of the matrix D are real, positive and single. There exists a matrix K such that

$$KDK^{-1} = \Lambda,$$

where

$$\Lambda = \begin{bmatrix} \lambda_1 & 0 \\ 0 & \lambda_2 \end{bmatrix}, \qquad K = \begin{bmatrix} 1 & k_{12} \\ k_{21} & 1 \end{bmatrix}, \tag{3}$$

with

$$k_{12} = \frac{\lambda_2 - d_{22}}{d_{21}}, \qquad k_{21} = \frac{\lambda_1 - d_{11}}{d_{12}}. \tag{4}$$

It is easy to verify that

$$k_{12} < 0, \qquad k_{21} > 0. \tag{5}$$

We shall deal with the following *free boundary value problem*:

Determine the functions $u_1 = u_1(x, t)$, $u_2 = u_2(x, t)$, $s = s(t)$ satisfying in the set

$$\Omega = \{(x, t): \ 0 < x < s(t), \ 0 < t < T < \infty\} \tag{6}$$

the parabolic system (1) and the boundary conditions

$$u_\alpha(x, 0) = \varphi_\alpha(x), \qquad 0 < x \leqq s(0) \equiv s_0, \quad \alpha = 1, 2, \tag{7}$$

$$u_1(0, t) = f(t), \qquad 0 < t < T, \tag{8}$$

$$u_1[s(t), t] = 0, \qquad 0 < t < T, \tag{9}$$

$$u_2[s(t), t] - u_2(0, t) = g(t), \qquad 0 < t < T, \tag{10}$$

$$\sum_{j=1}^{2} d_{1j} \frac{\partial u_j}{\partial x}\bigg|_{x=s(t)} = -s'(t), \qquad 0 < t < T, \tag{11}$$

$$\sum_{j=1}^{2} d_{2j} \frac{\partial u_j}{\partial x}\bigg|_{x=s(t)} = -\mu(t)\, u_2'(0, t), \qquad 0 < t < T, \tag{12}$$

where the functions f, φ_α, g, μ and the constant s_0 are given.

As a physical example of such a situation one can mention the problem of determining concentrations u_1, u_2 of two components in the three component one phase metal system, where one of the concentration is prescribed on the free boundary, while the second one satisfies some non local boundary condition (cf. [4], [5], [6], [9]). Conditions (11), (12) are related to the law of conservation of mass. The constants $d_{\alpha j}$ denote the diffusion coefficients.

The problem (1), (7)–(12) may be treated as a generalization of the one-dimensional one-phase Stefan problem considered by many authors, chiefly by A. Huber [3], A. Dacev [1], A. Friedman [2], L. Rubinstein [8] (see also the literature quoted in [8]). Our paper will be based on the results of A. Friedman [2]. We first transform our problem (1), (7)—(12) for the parabolic system into two separate problems for the heat equations and introduce Green functions for the half-plane. Thus the problem discussed in this paper is reduced to an equivalent system of non linear integral equations of Volterra type.

II. Definition of solution and list of assumptions

Let us denote

$$c(t) \equiv u_2(0, t) , \qquad c_0 \equiv u_2(0, 0) . \tag{13}$$

The set of functions $\{u_1, u_2, s\}$ will be called a *solution of the problem* (1), (7)—(12) if

1^0 the function $s(t)$ is positive and continuously differentiable in $\langle 0, T)$,

2^0 the functions $u_1(x, t)$, $u_2(x, t)$ are non-negative and continuous in $\bar{\Omega}$, and their derivatives $\frac{\partial u_\alpha}{\partial x}$, $\frac{\partial^2 u_\alpha}{\partial x^2}$, $\frac{\partial u_\alpha}{\partial t}$, $\alpha = 1, 2$, are continuous in Ω,

3^0 the function $c(t)$ is non-negative and continuously differentiable in $\langle 0, T)$, besides its derivative $c'(t)$ satisfies a Hölder condition and the inequality

$$c'(t) \leqq -k_{21} q_1(x, t) - q_2(x, t) - g'(t) \quad \text{in } \Omega , \tag{13'}$$

4^0 the equations (1), (7)—(12) are satisfied.

In order to solve the problem (1), (7)—(12) we admit the following assumptions:

I: The matrix D satisfies the conditions (2).

II: The functions $q_1(x, t)$, $q_2(x, t)$ are continuous in $\bar{\Omega}$ and they satisfy a Hölder condition with respect to the variable x in $\bar{\Omega}$; besides $q_\alpha(x, t) \leqq 0$ in $\bar{\Omega}$, $\alpha = 1, 2$.

III: The function $g(t)$ is continuously differentiable in $\langle 0, T)$, non-negative, and $g'(t)$ satisfies a Hölder condition.

IV: The function $f(t)$ is continuously differentiable in $\langle 0, T)$, non-negative and

$$k_{21} f(t) \geqq g(t) \geqq 0 , \qquad 0 \leqq t < T ,$$

where k_{21} is of the form (4) and satisfies the inequality (5).

V: The functions $\varphi_1(x)$, $\varphi_2(x)$ are continuously differentiable in $\langle 0, s_0\rangle$ and

$$\varphi_1(x) \geqq -k_{12}\varphi_2(x) \geqq 0 , \qquad 0 \leqq x \leqq s_0 ,$$

where k_{12} is of the form (4), and satisfies the inequality (5).

VI: The function $\mu(t)$ is defined and continuous in $\langle 0, T)$ and

$$\mu(t) \geqq \mu_0 > 0 , \qquad 0 \leqq t < T .$$

VII: There hold $f(0) = \varphi_1(0)$ and $u_2(0, 0) = \varphi_2(0) \equiv c_0$.

III. Integral equations equivalent to the problem (1), (7)—(12)

Let us introduce the vector W by

$$W = \begin{bmatrix} w_1 \\ w_2 \end{bmatrix} = \begin{bmatrix} 1 & k_{12} \\ k_{21} & 1 \end{bmatrix} \begin{bmatrix} u_1 \\ u_2 - c - g \end{bmatrix} \tag{14}$$

and transform the system (1), (7)—(12) into the following one:

$$\Lambda \frac{\partial^2 W}{\partial x^2} - \frac{\partial W}{\partial t} = K \left[\frac{dC}{dt} + Q\right] \quad \text{in } \Omega , \tag{15}$$

$$W(x, 0) = K\Phi , \qquad 0 < x \leqq s_0 , \tag{16}$$

$$W(0, t) = KF , \qquad 0 < t < T , \tag{17}$$

$$W[s(t), t] = \theta , \qquad 0 < t < T , \tag{18}$$

$$\Lambda \left.\frac{\partial W}{\partial x}\right|_{x=s(t)} = KS , \qquad 0 < t < T , \tag{19}$$

where we have denoted

$$\frac{dC}{dt} = \begin{bmatrix} 0 \\ \frac{dc}{dt} + \frac{dg}{dt} \end{bmatrix}, \qquad Q = \begin{bmatrix} q_1 \\ q_2 \end{bmatrix}, \qquad \theta = \begin{bmatrix} 0 \\ 0 \end{bmatrix},$$
$$\Phi = \begin{bmatrix} \varphi_1 \\ \varphi_2 - c_0 - g_0 \end{bmatrix}, \qquad F = \begin{bmatrix} f \\ -g \end{bmatrix}, \qquad S = \begin{bmatrix} -s' \\ -\mu c' \end{bmatrix}, \qquad g_0 = g(0) . \tag{20}$$

The problem (1), (7)–(12) is then divided into the two following ones:

Problem I:

$$\lambda_1 \frac{\partial^2 w_1}{\partial x^2} - \frac{\partial w_1}{\partial t} = k_{12}(c' + g') + q_1 + k_{12} q_2 \quad \text{in } \Omega\,, \tag{15'}$$

$$w_1(x, 0) = \varphi_1 + k_{12}(\varphi_2 - c_0 - g_0)\,, \qquad 0 < x \leqq s_0\,, \tag{16'}$$

$$w_1(0, t) = f - k_{12} g\,, \qquad 0 < t < T\,, \tag{17'}$$

$$w_1[s(t), t] = 0\,, \qquad 0 < t < T\,, \tag{18'}$$

$$\lambda_1 \frac{\partial w_1}{\partial x}\bigg|_{x=s(t)} = -s' - k_{12} c' \mu\,, \qquad 0 < t < T\,. \tag{19'}$$

Problem II:

$$\lambda_2 \frac{\partial^2 w_2}{\partial x^2} - \frac{\partial w_2}{\partial t} = c' + g' + k_{21} q_1 + q_2 \quad \text{in } \Omega\,, \tag{15''}$$

$$w_2(x, 0) = k_{21}\varphi_1 + \varphi_2 - c_0 - g_0\,, \qquad 0 < x \leqq s_0\,, \tag{16''}$$

$$w_2(0, t) = k_{21} f - g\,, \qquad 0 < t < T\,, \tag{17''}$$

$$w_2[s(t), t] = 0\,, \qquad 0 < t < T\,, \tag{18''}$$

$$\lambda_2 \frac{\partial w_2}{\partial x}\bigg|_{x=s(t)} = -k_{21} s' - \mu c'\,, \qquad 0 < t < T\,. \tag{19''}$$

From the conditions (19′) and (19″) we get

$$s' = \frac{k_{12}\lambda_2 z_2 - \lambda_1 z_1}{1 - k_{12}k_{21}}\,, \tag{21}$$

$$c' = -\frac{\lambda_2 z_2 - k_{21}\lambda_1 z_1}{\mu\,(1 - k_{12}k_{21})}\,, \tag{22}$$

where we have denoted

$$\frac{\partial w_1}{\partial x}\bigg|_{x=s(t)} = z_1\,, \qquad \frac{\partial w_2}{\partial x}\bigg|_{x=s(t)} = z_2\,, \qquad Z = \begin{bmatrix} z_1 \\ z_2 \end{bmatrix}. \tag{23}$$

Lemma: *If $\{w_1, w_2, s\}$ with corresponding c form a solution of the problems* I, II, *the assumptions* I–VII *are satisfied and the unknown function $c(t)$ satisfies the condition 3^0 of the definition, then the function $s(t)$ is monotone non-decreasing in t.*

Proof: According to the suppositions we have for $\alpha = 1, 2$:

$$w_\alpha(x, 0) \geqq 0\,, \qquad 0 < x < s_0\,, \qquad w_\alpha(0, t) \geqq 0\,, \qquad 0 < t < T\,,$$

$$\lambda_\alpha \frac{\partial^2 w_\alpha}{\partial x^2} - \frac{\partial w_\alpha}{\partial t} \leqq 0 \quad \text{in } \Omega\,, \qquad w_\alpha(x, t) \geqq 0 \quad \text{in } \Omega\,.$$

Hence owing to the maximum principle for parabolic equations [7]

$$\frac{\partial w_\alpha}{\partial x}\bigg|_{x=s(t)} \leqq 0$$

follows.

Now taking into account the equations (21) and (23), (5) we conclude that $s'(t) \geqq 0$.

Theorem 1: *If the assumptions* I–VII *are satisfied and the triplet $\{w_1, w_2, s\}$ with corresponding c forms a solution of the problems* I, II *for all $t \in (0, \sigma)$, then the functions $z_1(t)$, $z_2(t)$, defined by* (23) *satisfy the following system of nonlinear* VOLTERRA *equations*:

$$\begin{aligned} z_1(t) = {} & N_1[s(t), t; 0, 0]\,[\varphi_1(0) + k_{12}(\varphi_2(0) - c_0 - g_0) - f(0) + k_{12} g_0] - \\ & - 2\int_0^t N_1[s(t), t; 0, \tau]\left[\frac{\mathrm{d}f}{\mathrm{d}\tau} - k_{12}\frac{\mathrm{d}g}{\mathrm{d}\tau}\right]\mathrm{d}\tau + 2\int_0^t \frac{\partial}{\partial x} G_1[s(t), t; s(\tau), \tau]\, z_1(\tau)\,\mathrm{d}\tau + \\ & + 2\int_0^{s_0} N_1[s(t), t; \xi, 0]\left[\frac{\mathrm{d}\varphi_1}{\mathrm{d}\xi} + k_{12}\frac{\mathrm{d}\varphi_2}{\mathrm{d}\xi}\right]\mathrm{d}\xi + \int_0^t \int_0^{s(\tau)} \frac{\partial}{\partial x} G_1[s(t), t; \xi, \tau]\left[k_{12}\left(\frac{\mathrm{d}c}{\mathrm{d}\tau} + \frac{\mathrm{d}g}{\mathrm{d}\tau}\right) + \right. \\ & \left. + q_1 + k_{12} q_2\right]\mathrm{d}\xi\,\mathrm{d}\tau\,, \end{aligned} \tag{24}$$

$$z_2(t) = N_2[s(t), t; 0, 0]\,[k_{21}\varphi_1(0) + \varphi_2(0) - c_0 - k_{21}f(0)] -$$
$$- 2\int_0^t N_2[s(t), t; 0, 0]\left[k_{21}\frac{\mathrm{d}f}{\mathrm{d}\tau} - \frac{\mathrm{d}g}{\mathrm{d}\tau}\right]\mathrm{d}\tau + 2\int_0^t \frac{\mathrm{d}}{\mathrm{d}x} G_2[s(t), t; s(\tau), \tau]\, z_2(\tau)\,\mathrm{d}\tau +$$
$$+ 2\int_0^{s_0} N_2[s(t), t; \xi, 0]\left[k_{21}\frac{\mathrm{d}\varphi_1}{\mathrm{d}\xi} + \frac{\mathrm{d}\varphi_2}{\mathrm{d}\xi}\right]\mathrm{d}\xi + \int_0^t\int_0^{s(\tau)} \frac{\partial}{\partial x} G_2[s(t), t; \xi, \tau]\left[\frac{\partial c}{\partial \tau} + \frac{\mathrm{d}g}{\mathrm{d}\tau} + k_{21}q_1 + q_2\right]\mathrm{d}\xi\,\mathrm{d}\tau,$$

where

$$s(t) = s_0 + \int_0^t \frac{k_{12}\lambda_2 z_2 - \lambda_1 z_1}{1 - k_{12}k_{21}}\,\mathrm{d}\tau\,, \tag{25}$$

$$c(t) = c_0 - \int_0^t \frac{\lambda_2 z_2 - k_{21}\lambda_1 z_1}{\mu(1 - k_{12}k_{21})}\,\mathrm{d}\tau \tag{26}$$

and G_α, N_α, $\alpha = 1, 2$, denote the GREEN and NEUMANN functions for the half plane $x > 0$ related to the equations (15′) and (15″), respectively.

Proof: The GREEN functions G_α and NEUMANN functions N_α for the half plane $x > 0$ are given by:

$$G_\alpha(x, t; \xi, \tau) = \Gamma_\alpha(x, t; \xi, \tau) - \Gamma_\alpha(-x, t; \xi, \tau)\,, \qquad \alpha = 1, 2\,, \tag{27}$$
$$N_\alpha(x, t; \xi, \tau) = \Gamma_\alpha(x, t; \xi, \tau) + \Gamma_\alpha(-x, t; \xi, \tau)\,, \tag{28}$$

where

$$\Gamma_\alpha(x, t; \xi, \tau) = \frac{1}{\sqrt{2\pi(t - \tau)}}\exp\left[-\frac{(x - \xi)^2}{4\lambda_\alpha(t - \tau)}\right], \qquad \alpha = 1, 2\,.$$

Denote

$$\tilde{c}_1 = k_{12}\left(\frac{\mathrm{d}c}{\mathrm{d}t} + \frac{\mathrm{d}g}{\mathrm{d}t}\right) + q_1 + k_{12}q_2\,, \qquad \tilde{c}_2 = \frac{\mathrm{d}c}{\mathrm{d}t} + \frac{\mathrm{d}g}{\mathrm{d}t} + k_{21}q_1 + q_2 \tag{29}$$

and integrate the GREEN's identity

$$\frac{\partial}{\partial \xi}\left(G_\alpha\frac{\partial w_\alpha}{\partial \xi} - w_\alpha\frac{\partial G_\alpha}{\partial \xi}\right) + \frac{\partial}{\partial \tau}(G_\alpha w_\alpha) = G_\alpha \tilde{c}_\alpha\,, \qquad \alpha = 1, 2\,, \tag{30}$$

over the domain $0 < \xi < s(\tau)$, $0 < \varepsilon < \tau < t - \varepsilon$, letting $\varepsilon \to 0$. Taking into account the conditions (16), (17), (18) and denoting further

$$G = \begin{bmatrix} G_1 & 0 \\ 0 & G_2 \end{bmatrix}, \qquad N = \begin{bmatrix} N_1 & 0 \\ 0 & N_2 \end{bmatrix},$$

we get

$$W(x, t) = \int_0^t G[x, t; s(\tau), \tau]\frac{\partial W[s(\tau), \tau]}{\partial \xi}\,\mathrm{d}\tau + \int_0^t \frac{\partial}{\partial \xi} G[x, t; \xi, \tau]_{\xi=0}\,(KF)\,\mathrm{d}\tau +$$
$$+ \int_0^{s_0} G[x, t; \xi, 0]\,(K\Phi)\,\mathrm{d}\xi + \int_0^t\int_0^{s(\tau)} G[x, t; \xi, \tau]\left[K\left(\frac{\mathrm{d}C}{\mathrm{d}t} + Q\right)\right]\mathrm{d}\xi\,\mathrm{d}\tau\,. \tag{31}$$

Let us differentiate both sides of (31) with respect to x and let $x \to s(t) - 0$.

Using the known property concerning boundary values of the derivatives of the potential of simple layer and taking into account that $s(t)$ is non decreasing we obtain

$$Z(t) = 2N[s(t), t; 0, 0]\,[K(\Phi(0) - F(0))] - 2\int_0^t N[s(t), t; 0, \tau]\left(K\frac{\mathrm{d}F}{\mathrm{d}\tau}\right)\mathrm{d}\tau +$$
$$+ 2\int_0^t \frac{\partial}{\partial x} G[s(t), t; s(\tau), \tau]\, Z(\tau)\,\mathrm{d}\tau + 2\int_0^{s_0} N[s(t), t; \xi, 0]\left(K\frac{\mathrm{d}\Phi}{\mathrm{d}\xi}\right)\mathrm{d}\xi +$$
$$+ \int_0^t\int_0^{s(\tau)} \frac{\partial}{\partial x} G[s(t), t; \xi, \tau]\left[K\left(\frac{\mathrm{d}C}{\mathrm{d}\tau} + Q\right)\right]\mathrm{d}\xi\,\mathrm{d}\tau\,, \tag{32}$$

where $s(t)$ and $c(t)$ are defined by (25), (26), i.e. by equations (24).

Basing on the method of A. FRIEDMAN [2] we can now easily prove the following theorems:

Theorem II: *If*

1^0 *the assumptions* I—VII *are satisfied,*

2^0 $z_1(t)$, $z_2(t)$ *form a continuous solution of* (24) *for* $0 \leqq t < \sigma$, $\sigma > 0$ *with* $s(t)$, $c(t)$ *given by* (25), (26),

3^0 $s(t) > 0$, $c(t) \geqq 0$ *for* $0 \leqq t < \sigma$, (13') *is satisfied, then* $\{w_1, w_2, s\}$ *form a solution of the problems* I, II *for all* $t < \sigma$, *where* W *is defined by* (32) *and* $\partial W/\partial x$ *is replaced by* Z.

Theorem III: *The problem* (1), (7)—(12) *for* $0 \leqq t < \sigma$ *is equivalent to the problem of finding a continuous solution* $Z(t)$ *for the integral equation system* (32), *where* $s(t)$ and $c(t)$ *are given by* (25), (26), *are positive and* (13') *is satisfied; besides there is*

$$Z(t) = \left.\frac{\partial W}{\partial x}\right|_{x=s(t)} = K\left[\frac{\partial U}{\partial x}\right]_{x=s(t)}.$$

Thus the problem (1), (7)—(12) is reduced to the solution of a system of two nonlinear integral equations of VOLTERRA type. Proof of the existence and uniqueness of the solution to the system (32) with $s(t)$, $c(t)$ given by (25), (26) for $0 \leqq t < \sigma$ may be obtained by means of the BANACH fixpoint theorem. The form of equations (32) (or (24)) is similar to those obtained by A. FRIEDMAN [2], thus our proof would not essentially differ from the one given there.

References

1 Дацев, А., О линейной задаче Стефана, Доклады Акад. Наук СССР **58**, вып. 4, 563—566 (1947).
2 FRIEDMAN, A., Free boundary problems for parabolic equations. I: Melting of solids, J. Math. Mech. **8**, 499—517 (1959).
3 HUBER, A., Hauptaufsätze über das Fortschreiten der Schmelzgrenze in einem linearen Leiter, ZAMM **19**, 1—21 (1939).
4 KIRKALDY, J. S., Diffusion in multicomponent metallic systems, Canad. J. Phys. **35**, 435—440 (1957); **36**, 899—925 (1958).
5 Любов, Б. Я.; Плахотник, В. Т., О диффузионной кинетике роста интерметаллида при контакте слоев металлов, Физ. Металл. Металловед. **33**, вып. 2, 407—410 (1972).
6 MAJCHROWSKI, M., On construction of a solution of the parabolic system of partial differential equations. Demonstratio Math. **13**, 285—299 (1980).
7 NIRENBERG, L., A strong maximum principle for parabolic equations, Commun. Pure Appl. Math. **6**, 167—177 (1953).
8 RUBINSTEIN, L., The STEFAN Problem, A.M.S. Translations of Mathematical Monographs, Vol. 27, Providence, R.I., 1971. (Translation of Л. И. Рубинштейн, Проблема Стефана, издат. Звайгзне, Рига 1967).
Щербединский, Г. В.; Кондраченко, Л. А., Кинетика роста фаз в трехкомпонентных системах, Физ. Металл. Металловед. **33**, вып. 3, 487—494 (1972).

Eingereicht: 29. 9. 1980

Anschriften: Prof. Dr. JANINA WOLSKA-BOCHENEK, Technical University Warsaw, Institute of Mathematics, 00-661 Warsaw, Plac Jednosci Robotniczej 1, Poland; Prof. Dr. LOTHAR V. WOLFERSDORF, Mining Academy Freiberg, Department of Mathematics, 9200 Freiberg, Akademiestr. 6, GDR.

ZAMM 61, 547 – 557 (1981)

H. A. MANG / H. WALTER

Sturm-Sequence Check for Eigenproblems Resulting from Variational Principles with Subsidiary Conditions

Unter Benützung einer wohlbekannten Eigenschaft der Sturmschen Folgen läßt sich die Zahl der Eigenwerte, die kleiner als ein angenommener Wert ist, angeben. Der auf dieser Eigenschaft beruhende, sogenannte „Sturm-sequence check" stellt ein unentbehrliches Hilfsmittel für die Lösung von Eigenwertproblemen mit einer großen Anzahl modaler Koordinaten dar. Die Zielsetzung der vorliegenden Arbeit ist die Adaptierung des erwähnten „Sturm-sequence check" für die Anwendung auf Eigenwertprobleme, denen Variationsprinzipien mit Nebenbedingungen in der Form linearer Zwangsbedingungen für die Unbekannten zugrunde liegen, wie sie in der Ingenieurpraxis häufig auftreten. Der Aufsatz umfaßt eine theoretische Studie und eine darauf aufbauende numerische Untersuchung, in deren Rahmen (a) die theoretischen Erkenntnisse im Wege überschaubarer Handrechnung überprüft werden und (b) ein Computeralgorithmus zum Einsatz gelangt, der den modifizierten „Sturm-sequence check" enthält.

Permitting determination of the number of eigenvalues smaller than a shift to the right of the largest one of these eigenvalues, the well-known "Sturm-sequence check" is an indispensable remedy for the numerical solution of large-scale eigenproblems. The purpose of the present paper is to adapt the classical "Sturm-sequence check" for application to eigenproblems derived from variational principles with subsidiary conditions in the form of linear constraint conditions for the unknowns, as are frequently occurring in technical applications. The scope of this article encompasses theoretical considerations as well as a numerical study. The objectives of the numerical investigation are twofold: (a) numerical verification of the theoretical findings by means of transparent hand-calculation and (b) application of a computer algorithm containing the modified "Sturm-sequence check".

При решении больших проблем собственных значений полезно использовать известное свойство последовательностей Штурма, что с их помощью можно определить число собственных значений, меньших чем заданное значение. Целью настоящей работы является перенос использования указанного свойства на проблемы собственных значений, полученные из вариационных задач с побочными условиями в виде линейных ограничений на неизвестные, часто возникающими в технических приложениях. — Статья содержит и теоретическое исследование и численные примеры, в которых проверяются теоретические результаты с помощью ручного счета и, дальше, применяется программа для ЭВМ, написанная на основе рассмотренного подхода.

1. Introduction

Many eigenproblems encountered in technical applications are of a size such that solution for even only a few eigenpairs requires a significant computational effort. Fortunately, it is often sufficient to restrict treatment of a practical eigenproblem to extraction of just the smallest eigenvalues and computation of the corresponding eigenvectors.

In order to extract a certain eigenvalue by successively reducing the distance between lower and upper bounds to this eigenvalue, the so-called *Sturm-sequence check* [1] may be applied. This check is based on a count of the negative terms on the main-diagonal of the triangularized characteristic matrix of the generalized eigenproblem. For eigenproblems resulting from classical minimum principles, the number of the aforementioned negative diagonal terms is equal to the number of eigenvalues smaller than a shift to the right of the largest one of these eigenvalues. The Sturm-sequence check represents an important ingredient of the two most successful algorithms for the solution of large eigenproblems, the determinant-search technique and the subspace iteration method [1].

The purpose of the present paper is to adapt the Sturm-sequence check to eigenproblems derived from variational principles with subsidiary conditions in the form of linear constraint conditions for the unknowns. Typical representatives of such variational principles are modified Hamilton principles representing basic postulates, permitting construction of the mechanics of conservative systems subjected to subsidiary conditions of the aforementioned form.

The scope of this article encompasses theoretical considerations as well as a numerical study. A modified Hamilton principle serves as the starting point for the theoretical investigation. Discretization of the solid continuum under consideration is accomplished by means of finite elements. The motivation for employing the finite element method (FEM) as analysis tool for the numerical investigation rests with the fact that, more recently, a remarkable number of finite element techniques has been derived from variational principles with subsidiary conditions.

2. Theoretical considerations

At first, basic definitions concerning Sturm sequences, the eigenvalue separation property and the classical Sturm-sequence check are given. Thereafter, the generalized eigenproblem for a solid continuum subjected to free, undamped vibrations is formulated. Special features of the formulation are the FE-discretization of the continuum and specification of linear constraint conditions for the unknowns. A linear transformation of modal coordinates provides the basis for the adaptation of the Sturm-sequence check to eigenproblems of the initially mentioned class.

2.1 Basic definitions

A sequence of polynomials $p_i(x)$, $i = 1, 2, \dots, q$, forms a *Sturm sequence* if the roots of the polynomial $p_{j+1}(x)$ separate the roots of the polynomial $p_j(x)$ [1].

Consider the generalized eigenproblem

$$Av = \varrho Bv\,, \tag{1}$$

where A and B are two square-symmetric matrices of order n. The i-th nontrivial solution of equation (1) consists of the eigenvalue ϱ_i and the corresponding eigenvector v_i. Let $A^{(l)}$ and $B^{(l)}$ be square-symmetric matrices of order $n - l$, obtained by omitting the last l rows and columns of $A = A^{(0)}$ and $B = B^{(0)}$. It can be shown [1] that the characteristic polynomials of

$$A^{(l)}v^{(l)} = \varrho^{(l)}B^{(l)}v^{(l)}\,, \qquad l = 0, 1, \dots, n-1\,, \tag{2}$$

form a Sturm sequence, that is,

$$\varrho_1^{(l)} \leqq \varrho_1^{(l+1)} \leqq \varrho_2^{(l)} \leqq \varrho_2^{(l+1)} \leqq \dots \leqq \varrho_{n-l-1}^{(l)} \leqq \varrho_{n-l-1}^{(l+1)} \leqq \varrho_{n-l}^{(l)}\,, \qquad l = 0, 1, \dots, n-2\,. \tag{3}$$

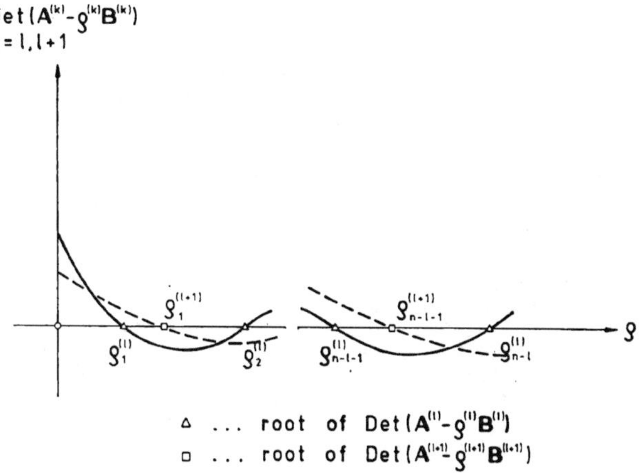

Fig. 1. Eigenvalue separation property

The above relationship represents the mathematical formulation of the eigenvalue separation property which is illustrated in Fig. 1.

If A is a positive definite matrix, the Sturm-sequence property yields that in the Gauss-factorization of $A - \mu B$ into LDL^T, where

$$L = \begin{bmatrix} 1 & & & & \\ l_{21} & 1 & & & \\ \vdots & & \ddots & & \\ \vdots & & & \ddots & \\ l_{n1} & l_{n2} & \cdots & l_{n,n-1} & 1 \end{bmatrix} \tag{4}$$

is a lower triangular matrix and

$$D = \begin{bmatrix} d_{11} & & & \\ & d_{22} & & \\ & & \ddots & \\ & & & d_{nn} \end{bmatrix} \tag{5}$$

is a diagonal matrix, the number of negative elements in D is equal to the number of eigenvalues smaller than μ, where μ represents a shift to the right of the largest one of these eigenvalues [1].

2.2 Generalized eigenproblem

Consider a linear elastic solid performing free, undamped vibrations. Fig. 2 illustrates the deformed position of this solid at a certain instant of time, t. The solid body is subdivided into s finite elements, two of which — the neighboring elements e and f — are shown in Fig. 2.

If the selected element displacement functions do not satisfy all essential boundary and interelement continuity conditions, suitable linear constraint conditions

$$Gq = 0 \tag{6}$$

must be formulated for the unknown nodal displacements q_i, $i = 1, 2, \dots, n$, listed as vector q [2]. Let m be the total number of such constraint equations. Then, G is an $(m \times n)$-matrix.

In the given case, a modification of Hamilton's principle will serve as the starting point for the formulation of the generalized eigenproblem [3]. This integral principle is given as

$$\int_{t_1}^{t_2} (T - U + C)\, \mathrm{d}t \to \text{Stationary Value}\,, \tag{7}$$

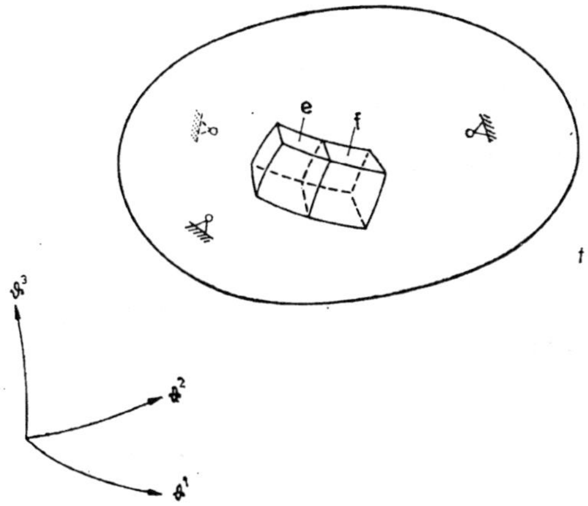

Fig. 2. Vibrating linear elastic solid at time t

wherein

$$T = \sum_{e=1}^{s} \int_{V^e} \tfrac{1}{2}\, \dot{\boldsymbol{u}}^{\mathrm{T}} \varrho \dot{\boldsymbol{u}}\, \mathrm{d}V\,, \tag{8}$$

is the kinetic energy,

$$U = \sum_{e=1}^{s} \int_{V^e} \tfrac{1}{2}\, \boldsymbol{\varepsilon}^{\mathrm{T}} \boldsymbol{\sigma}\, \mathrm{d}V\,, \tag{9}$$

is the strain energy and

$$C = \sum_{i=1}^{r} \int_{S^i} \boldsymbol{t}_d^{\mathrm{T}} \boldsymbol{d}\, \mathrm{d}S\,, \tag{10}$$

is a supplementary functional, permitting derivation of constraint conditions of the form of equation (6). Here, and in the following, the superscript "T" stands for "transpose".

The time integral extends from time t_1 to time t_2. The integrals in (8) and (9) stretch over the volumes V^e, $e = 1, 2, \dots, s$, of the finite elements. The integrals in equation (10) extend over those element interfaces S^i, $i = 1, 2, \dots, r$, at which certain essential interelement continuity conditions are not satisfied a priori.

The expression for the kinetic energy contains the derivative of the displacement vector $\boldsymbol{u}$ with respect to time; ϱ is the specific density. The expression for the strain energy contains the components of the linearized strain tensor, listed as vector $\boldsymbol{\varepsilon}$, and the components of the stress tensor, listed as vector $\boldsymbol{\sigma}$. The components of the vector $\boldsymbol{d}$, appearing in (10), represent discontinuities of displacement components at the respective element interface; $\boldsymbol{t}_d$ is a vector of corresponding forces which must not be expressed in terms of nodal displacements [2].

Expressing $\boldsymbol{u}$, $\boldsymbol{\varepsilon}$, $\boldsymbol{\sigma}$ and $\boldsymbol{d}$ in terms of $\boldsymbol{q}$ yields

$$\boldsymbol{u} = \boldsymbol{N}\boldsymbol{q}\,, \qquad \boldsymbol{\varepsilon} = \boldsymbol{N}_\varepsilon \boldsymbol{q}\,, \tag{11), (12}$$

and

$$\boldsymbol{\sigma} = \boldsymbol{E}\boldsymbol{\varepsilon} = \boldsymbol{E}\boldsymbol{N}_\varepsilon \boldsymbol{q}\,, \qquad \boldsymbol{d} = \boldsymbol{N}_d \boldsymbol{q}\,, \tag{13), (14}$$

where $\boldsymbol{N}$ is a matrix of shape functions and $\boldsymbol{E}$ is the elasticity matrix. Knowledge of $\boldsymbol{N}$ permits computation of $\boldsymbol{N}_\varepsilon$ and $\boldsymbol{N}_d$. Expressing $\boldsymbol{t}_d$ in terms of Lagrangian multipliers $\boldsymbol{\lambda}$, gives

$$\boldsymbol{t}_d = \boldsymbol{Q}\boldsymbol{\lambda}\,, \tag{15}$$

where $\boldsymbol{Q}$ represents a matrix of suitable basis functions.

Substituting equation (11) into (8) yields

$$T = \sum_{e=1}^{s} \tfrac{1}{2}\, \dot{\boldsymbol{q}}^{\mathrm{T}} \left(\int_{V^e} \boldsymbol{N}^{\mathrm{T}} \varrho \boldsymbol{N}\, \mathrm{d}V \right) \dot{\boldsymbol{q}}\,. \tag{16}$$

Substituting equations (12) and (13) into (9) gives

$$U = \sum_{e=1}^{s} \tfrac{1}{2}\, \boldsymbol{q}^{\mathrm{T}} \left(\int_{V^e} \boldsymbol{N}_\varepsilon^{\mathrm{T}} \boldsymbol{E} \boldsymbol{N}_\varepsilon\, \mathrm{d}V \right) \boldsymbol{q}\,. \tag{17}$$

Inserting equations (14) and (15) into (10) results in

$$C = \sum_{i=1}^{r} \boldsymbol{\lambda}^{\mathrm{T}} \left(\int_{S^i} \boldsymbol{Q}^{\mathrm{T}} \boldsymbol{N}_d\, \mathrm{d}S \right) \boldsymbol{q}\,. \tag{18}$$

For the sake of conciseness of the formulation, the abbreviations

$$\boldsymbol{M} = \sum_{e=1}^{s} \int_{V^e} \boldsymbol{N}^{\mathrm{T}} \varrho \boldsymbol{N}\, \mathrm{d}V\,, \qquad \boldsymbol{K} = \sum_{e=1}^{s} \int_{V^e} \boldsymbol{N}_\varepsilon^{\mathrm{T}} \boldsymbol{E} \boldsymbol{N}_\varepsilon\, \mathrm{d}V\,, \qquad \boldsymbol{G} = -\sum_{i=1}^{r} \int_{S^i} \boldsymbol{Q}^{\mathrm{T}} \boldsymbol{N}_d\, \mathrm{d}S \tag{19), (20), (21}$$

are introduced. $\boldsymbol{M}$ is the global mass matrix, $\boldsymbol{K}$ is the global stiffness matrix and $\boldsymbol{G}$ is the coefficient matrix appearing in the aforementioned set of linear constraint conditions [see equation (6)]. With the help of these system matrices the modified Hamilton principle can be rewritten as

$$\int_{t_1}^{t_2} (\tfrac{1}{2}\,\dot{\boldsymbol{q}}^{\mathsf T}\boldsymbol{M}\dot{\boldsymbol{q}} - \tfrac{1}{2}\,\boldsymbol{q}^{\mathsf T}\boldsymbol{K}\boldsymbol{q} - \boldsymbol{\lambda}^{\mathsf T}\boldsymbol{G}\boldsymbol{q})\,\mathrm{d}t \to \text{Stationary Value}\,. \tag{22}$$

For this functional to take on a stationary value, its first variation must vanish, that is,

$$\delta \int_{t_1}^{t_2} (\tfrac{1}{2}\,\dot{\boldsymbol{q}}^{\mathsf T}\boldsymbol{M}\dot{\boldsymbol{q}} - \tfrac{1}{2}\,\boldsymbol{q}^{\mathsf T}\boldsymbol{K}\boldsymbol{q} - \boldsymbol{\lambda}^{\mathsf T}\boldsymbol{G}\boldsymbol{q})\,\mathrm{d}t = 0\,. \tag{23}$$

Variation of the first term of the above integral yields

$$\delta \int_{t_1}^{t_2} \tfrac{1}{2}\,\dot{\boldsymbol{q}}^{\mathsf T}\boldsymbol{M}\dot{\boldsymbol{q}}\,\mathrm{d}t = \int_{t_1}^{t_2} \delta\dot{\boldsymbol{q}}^{\mathsf T}\boldsymbol{M}\dot{\boldsymbol{q}}\,\mathrm{d}t = \delta\boldsymbol{q}^{\mathsf T}\boldsymbol{M}\dot{\boldsymbol{q}}\big|_{t_1}^{t_2} - \int_{t_1}^{t_2} \delta\boldsymbol{q}^{\mathsf T}\boldsymbol{M}\ddot{\boldsymbol{q}}\,\mathrm{d}t\,. \tag{24}$$

Note that because of

$$\delta\boldsymbol{q}|_{t_1} = \delta\boldsymbol{q}|_{t_2} = \boldsymbol{0}\,, \qquad \delta\boldsymbol{q}^{\mathsf T}\boldsymbol{M}\dot{\boldsymbol{q}}\big|_{t_1}^{t_2} = 0\,. \tag{25), (26}$$

Making use of equations (24) and (26) and performing the two remaining variations in (23) results in

$$\int_{t_1}^{t_2} [\delta\boldsymbol{q}^{\mathsf T}(\boldsymbol{M}\ddot{\boldsymbol{q}} + \boldsymbol{K}\boldsymbol{q} + \boldsymbol{G}^{\mathsf T}\boldsymbol{\lambda}) + \delta\boldsymbol{\lambda}^{\mathsf T}\boldsymbol{G}\boldsymbol{q}]\,\mathrm{d}t = 0\,. \tag{27}$$

For the above integral to vanish for arbitrary variations, its integrand must vanish at each instant of time, that is,

$$\delta\boldsymbol{q}^{\mathsf T}(\boldsymbol{M}\ddot{\boldsymbol{q}} + \boldsymbol{K}\boldsymbol{q} + \boldsymbol{G}^{\mathsf T}\boldsymbol{\lambda}) + \delta\boldsymbol{\lambda}^{\mathsf T}\boldsymbol{G}\boldsymbol{q} = 0\,. \tag{28}$$

Free vibrations involve harmonic motion in each mode shape. Choosing

$$\boldsymbol{q} = \hat{\boldsymbol{q}}\sin\omega t \tag{29.1}$$

it follows by physical reasons that

$$\boldsymbol{\lambda} = \hat{\boldsymbol{\lambda}}\sin\omega t\,. \tag{29.2}$$

Substitution of (29.1) and (29.2) into equation (28) yields

$$\delta\hat{\boldsymbol{q}}^{\mathsf T}[(\boldsymbol{K} - \omega^2\boldsymbol{M})\,\hat{\boldsymbol{q}} + \boldsymbol{G}^{\mathsf T}\hat{\boldsymbol{\lambda}}] + \delta\hat{\boldsymbol{\lambda}}^{\mathsf T}\boldsymbol{G}\hat{\boldsymbol{q}} = 0\,, \tag{30}$$

where $\hat{\boldsymbol{q}}$ and $\hat{\boldsymbol{\lambda}}$ denote vectors of modal coordinates and ω is the circular frequency. Equation (30) represents the basis for the solution of the generalized eigenproblem.

2.3 Transformation of modal coordinates

For the purpose of demonstrating how the Sturm-sequence check can be adapted to eigenproblems derived from variational principles with subsidiary conditions of the previously mentioned form, it has proved useful to conduct the following transformation of modal coordinates:

$$\hat{\boldsymbol{q}}^* = \begin{Bmatrix} \hat{\boldsymbol{q}}_1 \\ \hat{\boldsymbol{r}} \end{Bmatrix} = \begin{bmatrix} \boldsymbol{I}^{(n-m)} & \boldsymbol{0} \\ \boldsymbol{G}_1 & \boldsymbol{G}_2 \end{bmatrix} \begin{Bmatrix} \hat{\boldsymbol{q}}_1 \\ \hat{\boldsymbol{q}}_2 \end{Bmatrix} = \begin{Bmatrix} \hat{\boldsymbol{q}}_1 \\ \boldsymbol{G}\hat{\boldsymbol{q}} \end{Bmatrix}. \tag{31}$$

Note that because of the existence of m constraint conditions for the n nodal displacements, m modal coordinates, listed as vector $\hat{\boldsymbol{q}}_2$, depend linearly on the remaining $n - m$ modal coordinates, listed as vector $\hat{\boldsymbol{q}}_1$. In (31), $\boldsymbol{I}^{(n-m)}$ is a $[(n - m) \times (n - m)]$-unit matrix; $\boldsymbol{G}_1$ and $\boldsymbol{G}_2$ are submatrices of $\boldsymbol{G}$ of orders $[m \times (n - m)]$ and $(m \times m)$, respectively.

On the basis of this transformation, equation (30) can be rewritten as

$$\delta\hat{\boldsymbol{q}}^{*\mathsf T}(\boldsymbol{K}^* - \omega^2\boldsymbol{M}^*)\,\hat{\boldsymbol{q}}^* + \delta\hat{\boldsymbol{r}}^{\mathsf T}\hat{\boldsymbol{\lambda}} + \delta\hat{\boldsymbol{\lambda}}^{\mathsf T}\hat{\boldsymbol{r}} = 0\,, \tag{32}$$

where

$$\delta\hat{\boldsymbol{q}}^{*\mathsf T}(\boldsymbol{K}^* - \omega^2\boldsymbol{M}^*)\,\hat{\boldsymbol{q}}^* = \delta\hat{\boldsymbol{q}}^{\mathsf T}(\boldsymbol{K} - \omega^2\boldsymbol{M})\,\hat{\boldsymbol{q}}\,. \tag{33}$$

With the help of

$$\hat{\boldsymbol{q}}_2 = \boldsymbol{G}_2^{-1}(\hat{\boldsymbol{r}} - \boldsymbol{G}_1\hat{\boldsymbol{q}}_1)\,, \tag{34}$$

resulting from equation (31), and with the abbreviations

$$\boldsymbol{H} = \boldsymbol{K} - \omega^2\boldsymbol{M}\,, \qquad \boldsymbol{H}^* = \boldsymbol{K}^* - \omega^2\boldsymbol{M}^*\,, \tag{35.1), (35.2}$$

equation (33) can be rewritten as

$$\lfloor \delta\hat{\boldsymbol{q}}_1^{\mathsf T}\ \delta\hat{\boldsymbol{r}}^{\mathsf T} \rfloor \begin{bmatrix} \boldsymbol{H}_{11}^* & \boldsymbol{H}_{12}^* \\ \boldsymbol{H}_{12}^{*\mathsf T} & \boldsymbol{H}_{22}^* \end{bmatrix} \begin{Bmatrix} \hat{\boldsymbol{q}}_1 \\ \hat{\boldsymbol{r}} \end{Bmatrix} = \lfloor \delta\hat{\boldsymbol{q}}_1^{\mathsf T}\ (\delta\hat{\boldsymbol{r}}^{\mathsf T} - \delta\hat{\boldsymbol{q}}_1^{\mathsf T}\boldsymbol{G}_1^{\mathsf T})\,(\boldsymbol{G}_2^{-1})^{\mathsf T} \rfloor \begin{bmatrix} \boldsymbol{H}_{11} & \boldsymbol{H}_{12} \\ \boldsymbol{H}_{12}^{\mathsf T} & \boldsymbol{H}_{22} \end{bmatrix} \begin{Bmatrix} \hat{\boldsymbol{q}}_1 \\ \boldsymbol{G}_2^{-1}(\hat{\boldsymbol{r}} - \boldsymbol{G}_1\hat{\boldsymbol{q}}_1) \end{Bmatrix}, \tag{36}$$

where $\boldsymbol{H}_{ij}^*$, $i, j = 1, 2$, and $\boldsymbol{H}_{kl}$, $k, l = 1, 2$, are submatrices of $\boldsymbol{H}^*$ and $\boldsymbol{H}$, respectively.

Comparing coefficients of equation (36) yields

$$\boldsymbol{H}_{11}^* = \boldsymbol{H}_{11} - \boldsymbol{G}_1^{\mathsf T}(\boldsymbol{G}_2^{-1})^{\mathsf T}\,\boldsymbol{H}_{12}^{\mathsf T} - \boldsymbol{H}_{12}\boldsymbol{G}_2^{-1}\boldsymbol{G}_1 + \boldsymbol{G}_1^{\mathsf T}(\boldsymbol{G}_2^{-1})^{\mathsf T}\,\boldsymbol{H}_{22}\boldsymbol{G}_2^{-1}\boldsymbol{G}_1\,, \tag{37.1}$$

$$\boldsymbol{H}_{12}^* = \boldsymbol{H}_{12}\boldsymbol{G}_2^{-1} - \boldsymbol{G}_1^{\mathsf T}(\boldsymbol{G}_2^{-1})^{\mathsf T}\,\boldsymbol{H}_{22}\boldsymbol{G}_2^{-1}\,, \qquad \boldsymbol{H}_{22}^* = (\boldsymbol{G}_2^{-1})^{\mathsf T}\,\boldsymbol{H}_{22}\boldsymbol{G}_2^{-1}\,. \tag{37.2), (37.3}$$

With the help of the expression on the left side of equation (36), equation (32) can be rewritten as follows:

$$\lfloor \delta\hat{q}_1^{\mathsf T} \vdots \delta\hat{r}^{\mathsf T} \vdots \delta\lambda^{\mathsf T} \rfloor \begin{bmatrix} H_{11}^* & H_{12}^* & 0 \\ H_{12}^{*\mathsf T} & H_{22}^* & I^{(m)} \\ 0 & I^{(m)} & 0 \end{bmatrix} \begin{Bmatrix} \hat{q}_1 \\ \hat{r} \\ \hat{\lambda} \end{Bmatrix} = 0 . \tag{38}$$

In equation (38), $I^{(m)}$ denotes an $(m \times m)$-unit matrix. Equation (38) is satisfied for arbitrary variations $\delta\hat{q}_1$ and $\delta\hat{\lambda}$ if

$$\begin{bmatrix} H_{11}^* & H_{12}^* & 0 \\ H_{12}^{*\mathsf T} & H_{22}^* & I^{(m)} \\ 0 & I^{(m)} & 0 \end{bmatrix} \begin{Bmatrix} \hat{q}_1 \\ \hat{r} \\ \hat{\lambda} \end{Bmatrix} = \begin{Bmatrix} 0 \\ 0 \\ 0 \end{Bmatrix}. \tag{39}$$

The above system of homogeneous algebraic equations has a nontrivial solution if

$$\operatorname{Det} \begin{bmatrix} H_{11}^* & H_{12}^* & 0 \\ H_{12}^{*\mathsf T} & H_{22}^* & I^{(m)} \\ 0 & I^{(m)} & 0 \end{bmatrix} = 0 . \tag{40}$$

Because of

$$\begin{bmatrix} H_{11}^* & H_{12}^* & 0 \\ H_{12}^{*\mathsf T} & H_{22}^* & I^{(m)} \\ 0 & I^{(m)} & 0 \end{bmatrix} = \begin{bmatrix} I^{(n-m)} & -G_1^{\mathsf T}(G_2^{-1})^{\mathsf T} & 0 \\ 0 & (G_2^{-1})^{\mathsf T} & 0 \\ 0 & 0 & I^{(m)} \end{bmatrix} \begin{bmatrix} H_{11} & H_{12} & G_1^{\mathsf T} \\ H_{12}^{\mathsf T} & H_{22} & G_2^{\mathsf T} \\ G_1 & G_2 & 0 \end{bmatrix} \begin{bmatrix} I^{(n-m)} & 0 & 0 \\ -G_2^{-1}G_1 & G_2^{-1} & 0 \\ 0 & 0 & I^{(m)} \end{bmatrix}, \tag{41}$$

there is

$$\operatorname{Det} \begin{bmatrix} H_{11}^* & H_{12}^* & 0 \\ H_{12}^{*\mathsf T} & H_{22}^* & I^{(m)} \\ 0 & I^{(m)} & 0 \end{bmatrix} = c \operatorname{Det} \begin{bmatrix} H_{11} & H_{12} & G_1^{\mathsf T} \\ H_{12}^{\mathsf T} & H_{22} & G_2^{\mathsf T} \\ G_1 & G_2 & 0 \end{bmatrix}, \tag{42}$$

where

$$c = \operatorname{Det} \begin{bmatrix} I^{(n-m)} & -G_1^{\mathsf T}(G_2^{-1})^{\mathsf T} & 0 \\ 0 & (G_2^{-1})^{\mathsf T} & 0 \\ 0 & 0 & I^{(m)} \end{bmatrix} \cdot \operatorname{Det} \begin{bmatrix} I^{(n-m)} & 0 & 0 \\ -G_2^{-1}G_1 & G_2^{-1} & 0 \\ 0 & 0 & I^{(m)} \end{bmatrix} > 0 . \tag{43}$$

The characteristic polynomial on the left side of equation (42) differs from the characteristic polynomial on the right side of this equation only by the factor c, defined in (43). Consequently, the roots of both polynomials, that is, the eigenvalues of the two characteristic matrices, are identical.

2.4 Sturm-sequence check

Since c is a positive multiplier, application of the Sturm-sequence check to the coefficient matrix on the left side of equation (42) must result in the same number of negative terms on the main-diagonal of the triangularized matrix as is obtained when applying the Sturm-sequence check to the original characteristic matrix, shown on the right side of equation (42).

Partial Gauss-elimination of the coefficient matrix on the left side of (42) yields

$$R_{(n-m+1)} = \left[\begin{array}{cccc:cccc:c} d_{11} & l_{21} & \cdots & l_{n-m,1} & l_{n-m+1,1} & l_{n-m+2,1} & \cdots & l_{n1} & \\ & d_{22} & \cdots & l_{n-m,2} & l_{n-m+1,2} & l_{n-m+2,2} & \cdots & l_{n2} & 0 \\ & & \ddots & \vdots & \vdots & \vdots & & \vdots & \\ & 0 & & d_{n-m,n-m} & l_{n-m+1,n-m} & l_{n-m+2,n-m} & \cdots & l_{n,n-m} & \\ \hdashline & & & & d_{n-m+1,n-m+1} & l_{n-m+2,n-m+1} & \cdots & l_{n,n-m+1} & \\ & 0 & & & l_{n-m+2,n-m+1} & g_{n-m+2,n-m+2} & \cdots & g_{n,n-m+2} & I^{(m)} \\ & & & & \vdots & \vdots & & \vdots & \\ & & & & l_{n,n-m+1} & g_{n,n-m+2} & \cdots\cdots & g_{nn} & \\ \hdashline & 0 & & & & I^{(m)} & & & 0 \end{array}\right] \tag{44}$$

where d_{ii}, $i = 1, 2, \ldots, n - m + 1$, and l_{jk}, $j > k$, $j = 2, 3, \ldots, n$, $k = 1, 2, \ldots, n - m + 1$, are coefficients of the diagonal matrix $\boldsymbol{D}$ and the lower triangular matrix $\boldsymbol{L}$, respectively, contained in the $\boldsymbol{LDL}^\mathsf{T}$ decomposition of

$$\begin{bmatrix} \boldsymbol{H}_{11}^* & \boldsymbol{H}_{12}^* \\ \boldsymbol{H}_{12}^{*\mathsf{T}} & \boldsymbol{H}_{22}^* \end{bmatrix}.$$

Matrix coefficients $g_{rs} = g_{sr}$, $r, s = n - m + 2, n - m + 3, \ldots, n$, in equation (44) have replaced corresponding coefficients of the submatrix $\boldsymbol{H}_{22}^*$. Employing the unit matrix in the third row of submatrices of $\boldsymbol{R}_{(n-m+1)}$ for reduction of the central submatrix of $\boldsymbol{R}_{(n-m+1)}$ to a diagonal matrix which, thereafter, is used for reduction of the aforementioned unit matrix to a null matrix, one finally obtains the following upper triangular matrix:

$$\boldsymbol{U} = \left[\begin{array}{cccc:cccc:cccc}
d_{11} & l_{21} & \cdots & l_{n-m,1} & l_{n-m+1,1} & l_{n-m+2,1} & \cdots & l_{n1} & & & & \\
 & d_{22} & \cdots & l_{n-m,2} & l_{n-m+1,2} & l_{n-m+2,2} & \cdots & l_{n2} & & \boldsymbol{0} & & \\
 & & \ddots & \vdots & \vdots & \vdots & & \vdots & & & & \\
\boldsymbol{0} & & & d_{n-m,n-m} & l_{n-m+1,n-m} & l_{n-m+2,n-m} & \cdots & l_{n,n-m} & & & & \\
\hdashline
 & & & & d_{n-m+1,n-m+1} & & & \boldsymbol{0} & & & & \\
 & \boldsymbol{0} & & & & g_{n-m+2,n-m+2} & & & & \boldsymbol{I}^{(m)} & & \\
 & & & & & & \ddots & & & & & \\
 & & & & \boldsymbol{0} & & & g_{nn} & & & & \\
\hdashline
 & & & & & & & & -\dfrac{1}{d_{n-m+1,n-m+1}} & & & \boldsymbol{0} \\
 & \boldsymbol{0} & & & & \boldsymbol{0} & & & & -\dfrac{1}{g_{n-m+2,n-m+2}} & & \\
 & & & & & & & & & & \ddots & \\
 & & & & & & & & \boldsymbol{0} & & & -\dfrac{1}{g_{nn}}
\end{array}\right]. \tag{45}$$

Multiplication of the terms located on the main-diagonal of $\boldsymbol{U}$ shows that

$$\operatorname{Det} \begin{bmatrix} \boldsymbol{H}_{11}^* & \boldsymbol{H}_{12}^* & \boldsymbol{0} \\ \boldsymbol{H}_{12}^{*\mathsf{T}} & \boldsymbol{H}_{22}^* & \boldsymbol{I}^{(m)} \\ \boldsymbol{0} & \boldsymbol{I}^{(m)} & \boldsymbol{0} \end{bmatrix} = (-1)^m \operatorname{Det} [\boldsymbol{H}_{11}^*]. \tag{46}$$

Thus, the eigenvalues of the three characteristic matrices

$$\begin{bmatrix} \boldsymbol{K} - \omega^2\boldsymbol{M} & \boldsymbol{G}^\mathsf{T} \\ \boldsymbol{G} & \boldsymbol{0} \end{bmatrix}, \quad \begin{bmatrix} \boldsymbol{H}_{11}^* & \boldsymbol{H}_{12}^* & \boldsymbol{0} \\ \boldsymbol{H}_{12}^{*\mathsf{T}} & \boldsymbol{H}_{22}^* & \boldsymbol{I}^{(m)} \\ \boldsymbol{0} & \boldsymbol{I}^{(m)} & \boldsymbol{0} \end{bmatrix}, \quad [\boldsymbol{H}_{11}^*]$$

must be identical.

In principle, the Sturm-sequence check could be applied directly to the matrix $\boldsymbol{H}_{11}^*$. Since the corresponding eigenvector, $\hat{\boldsymbol{q}}_1$, does not contain linearly dependent modal coordinates, no adaption of this check would be necessary. However, because of the need to perform lengthy matrix operations to obtain $\boldsymbol{H}_{11}^*$ [see equation (37.1)], it appears impracticable to apply the Sturm-sequence check directly to $\boldsymbol{H}_{11}^*$. Thus, the question arises, how to obtain the number of negative terms, k, on the main-diagonal of a matrix resulting from triangularization of $\boldsymbol{H}_{11}^*$ without actually computing $\boldsymbol{H}_{11}^*$. In the following, this question will be answered.

Assume that application of the Sturm-sequence check to the original characteristic matrix, that is, to the left one of the three above matrices, has indicated s negative terms on the main-diagonal of the respective triangularized matrix. Consequently, the main-diagonal of the matrix stemming from triangularization of the middle one of the three above matrices must also contain s negative signs. Since there are m negative terms among the last $2m$ terms on the main-diagonal of the matrix $\boldsymbol{U}$, stemming from the middle one of the three above matrices, it follows that

$$k = s - m. \tag{47}$$

Thus, for eigenproblems resulting from variational principles with subsidiary conditions of the initially introduced form, the number of constraint conditions, m, which is known a priori, must be subtracted from the total number of negative terms, s, on the main-diagonal of a matrix resulting from triangularization of the original characteristic matrix, in order to obtain the number of eigenvalues, k, which are smaller than a shift to the right of the k-th eigenvalue.

For practical calculations the employed characteristic matrix differs from the left one of the three above matrices in so far as the upper submatrix $[K - \omega^2 M \vdots G^\mathsf{T}]$ is interspersed with rows of the lower submatrix $[G \vdots 0]$.

3. Numerical investigation

The objectives of the numerical investigation are twofold:

(a) numerical verification of the theoretical findings by means of transparent "hand-calculations" and

(b) application of a computer algorithm containing the adapted Sturm-sequence check. The first objective will be met through computation of circular frequencies of a triangular slab. The problem of buckling of a narrow plate strip subjected to dead load provides the vehicle for meeting the second objective.

3.1 Circular frequencies of a triangular slab

Fig. 3 shows a triangular slab with given dimensions. The slab is assumed to be simply supported along its edges. Thickness, h, modulus of elasticity, E, Poisson's ratio, μ, and specific density, ϱ, are chosen as $h = 1$, $E = 12$, $\mu = 0$ and $\varrho = 5040$. Consistency of units is tacitly assumed

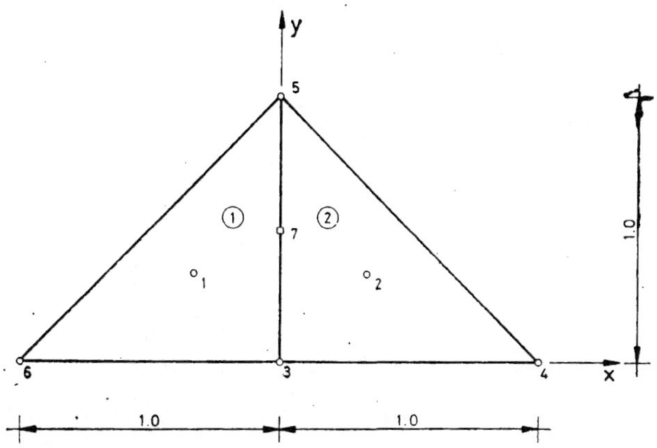

Fig. 3. Simply supported triangular slab divided into two finite elements

The discretized slab consists of two triangular finite elements. Details concerning the selected finite element are given, for example, in [2]. The rationale for choosing such an unrealistically coarse mesh is the aim to verify the theoretical findings without resorting to the digital computer. In view of this objective, deviations of numerical results from "true" solutions are irrelevant.

After satisfaction of the boundary conditions, all but the following three degrees of freedom (d.o.f.) are deactivated: w_1, w_2 and ϑ_{x3}, representing the transverse displacements at the element centroids 1 und 2, respectively, and the rotation about the x-axis at node point 3.

The stiffness matrix for the discretized slab is taken from Ref. [2]. It is given as

$$\boldsymbol{K} = \begin{bmatrix} 729 & 0 & -\frac{81}{2} \\ 0 & 729 & -\frac{81}{2} \\ -\frac{81}{2} & -\frac{81}{2} & 8 \end{bmatrix}. \tag{48}$$

The mass matrix is obtained as

$$\boldsymbol{M} = \begin{bmatrix} 729 & 0 & -\frac{27}{2} \\ 0 & 729 & \frac{27}{2} \\ -\frac{27}{3} & \frac{27}{2} & 8 \end{bmatrix}. \tag{49}$$

In order to enforce continuity of normal slopes (so called C^1-continuity) along the interelement boundary line $\overline{3\,5}$, the constraint condition

$$\vartheta_{y7}^{(1)} - \vartheta_{y7}^{(2)} = 0 \tag{50}$$

must be satisfied [2], where $\vartheta_{y7}^{(1)}$ and $\vartheta_{y7}^{(2)}$ denote element rotations about the y-axis at node point 7, located in the middle between node points 3 and 5. The superscripts in equation (50) refer to the elements. Expressing equation (49) in terms of nodal displacements yields [2]

$$\vartheta_{y7}^{(1)} - \vartheta_{y7}^{(2)} = \boldsymbol{G}\boldsymbol{q} = \left\lfloor -\frac{27}{4},\ -\frac{27}{4},\ \frac{3}{2} \right\rfloor \begin{Bmatrix} w_1 \\ w_2 \\ \vartheta_{x3} \end{Bmatrix} = 0\,. \tag{51}$$

Considering ϑ_{x3} as the linearly dependent d.o.f., results in

$$q_1^{\mathrm{T}} = \lfloor w_1 w_2 \rfloor , \qquad r = Gq = \vartheta_{y7}^{(1)} - \vartheta_{y7}^{(2)} , \tag{52.1), (52.2}$$

$$G_1 = \left\lfloor -\frac{27}{4}, \; -\frac{27}{4} \right\rfloor , \qquad G_2 = \left\lfloor \frac{3}{2} \right\rfloor . \tag{53.1), (53.2}$$

Note that, in the given case, r and G_2 degenerate to scalar quantities.

Transformation of

$$\left[\begin{array}{cc:c:c} H_{11} & H_{12} & G_1^{\mathrm{T}} \\ \hdashline H_{12}^{\mathrm{T}} & H_{22} & G_2^{\mathrm{T}} \\ \hdashline G_1 & G_2 & 0 \end{array}\right] = \left[\begin{array}{cc:c:c} 729(1-\omega^2) & 0 & \frac{27}{2}(-3+\omega^2) & -\frac{27}{4} \\ 0 & 729(1-\omega^2) & \frac{27}{2}(-3-\omega^2) & -\frac{27}{4} \\ \hdashline \frac{27}{2}(-3+\omega^2) & \frac{27}{2}(-3+\omega^2) & 8(1-\omega^2) & \frac{3}{2} \\ \hdashline -\frac{27}{4} & -\frac{27}{4} & \frac{3}{2} & 0 \end{array}\right] \tag{54}$$

according to equation (41) yields

$$\left[\begin{array}{cc:c:c} H_{11}^{*} & H_{12}^{*} & 0 \\ \hdashline H_{12}^{*\mathrm{T}} & H_{22}^{*} & I^{(m)} \\ \hdashline 0 & I^{(m)} & 0 \end{array}\right] = \left[\begin{array}{cc:c:c} \left(\frac{1053}{2} - \frac{1539}{2}\omega^2\right) & \left(-\frac{405}{2} - 162\omega^2\right) & -3-15\omega^2 & 0 \\ \left(-\frac{405}{2} - 162\omega^2\right) & \left(\frac{1053}{2} - \frac{2025}{2}\omega^2\right) & -3-33\omega^2 & 0 \\ \hdashline -3-15\omega^2 & -3-33\omega^2 & \frac{32}{9}(1-\omega^2) & 1 \\ \hdashline 0 & 0 & 1 & 0 \end{array}\right] . \tag{55}$$

Recalling that the eigenvalues of the above characteristic matrix must be identical to the eigenvalues of the submatrix $[H_{11}^{*}]$, the respective eigenvalues are computed from

$$\operatorname{Det}\,[H_{11}^{*}] = 51\omega^4 - 68\omega^2 + 16 = 0 \tag{56}$$

as

$$\omega_1^2 = 0{,}3051 , \qquad \omega_2^2 = 1{,}0282 . \tag{57}$$

In order to verify the "modified Sturm-sequence check" numerically, the three matrices beneath equation (46) are evaluated for a shift to the right of ω_2^2, chosen as $\bar{\omega}^2 = \frac{3}{2}$. This gives

$$\begin{bmatrix} -\frac{729}{2} & 0 & -\frac{81}{4} & -\frac{27}{4} \\ 0 & -\frac{729}{2} & -\frac{243}{4} & -\frac{27}{4} \\ -\frac{81}{4} & -\frac{243}{4} & -4 & \frac{3}{2} \\ -\frac{27}{4} & -\frac{27}{4} & \frac{3}{2} & 0 \end{bmatrix}, \quad \begin{bmatrix} -\frac{2511}{4} & -\frac{891}{2} & -\frac{51}{2} & 0 \\ -\frac{891}{2} & -\frac{3969}{4} & -\frac{105}{2} & 0 \\ -\frac{51}{2} & -\frac{105}{2} & -\frac{16}{9} & 1 \\ 0 & 0 & 1 & 0 \end{bmatrix}, \quad \begin{bmatrix} -\frac{2511}{4} & -\frac{891}{2} \\ -\frac{891}{2} & -\frac{3969}{4} \end{bmatrix}$$

Triangularization of these matrices yields

$$\begin{bmatrix} -\frac{729}{2} & 0 & -\frac{81}{4} & -\frac{27}{4} \\ & -\frac{729}{12} & -\frac{243}{4} & -\frac{27}{4} \\ & & \frac{29}{4} & 3 \\ 0 & & & -\frac{115}{116} \end{bmatrix}, \quad \begin{bmatrix} -\frac{2511}{4} & -\frac{891}{2} & -\frac{51}{2} & 0 \\ & -\frac{83835}{124} & -\frac{2133}{62} & 0 \\ & & \frac{116}{115} & 1 \\ 0 & & & -\frac{115}{116} \end{bmatrix}, \quad \begin{bmatrix} -\frac{2511}{4} & -\frac{891}{2} \\ 0 & -\frac{83835}{124} \end{bmatrix}.$$

The main-diagonal of the left one of the three above matrices, that is, of the triangularized original matrix as employed for practical computations, contains three negative signs. Its determinant is obtained as $(-3^{12}/2^6)$ 115. According to equation (43), the value of the determinant of the central matrix of the three above matrices can be obtained by multiplying the determinant of the left one of these matrices by

$$c = \mathrm{Det}\begin{bmatrix} 1 & 0 & \frac{9}{2} & 0 \\ 0 & 1 & \frac{9}{2} & 0 \\ 0 & 0 & \frac{2}{3} & 0 \\ 0 & 0 & 0 & 1 \end{bmatrix} \cdot \mathrm{Det}\begin{bmatrix} 1 & 0 & 0 & 0 \\ 0 & 1 & 0 & 0 \\ \frac{9}{2} & \frac{9}{2} & \frac{2}{3} & 0 \\ 0 & 0 & 0 & 1 \end{bmatrix} = \left(\frac{2}{3}\right)^2. \tag{58}$$

Noting that the determinant of the aforementioned central matrix is $(-3^{10}/2^4)$. 115, the validity of equation (43) is easily verified:

$$-\frac{3^{10}}{2^4}\,115 = \left(\frac{2}{3}\right)^2 \left(-\frac{3^{12}}{2^6}\right) 115\,. \tag{59}$$

Comparing the product of the terms on the main diagonal of the central matrix to the product of the terms on the main diagonal of the right one of the three described matrices, it is seen that the absolute values of both determinants are identical and that equation (46) with $m = 1$ is satisfied. Noting that the matrix on the left side has three negative signs on the main-diagonal, that is, $s = 3$, and that the matrix on the right side has two negative signs on the main-diagonal, that is, $k = 2$, and recalling that one constraint condition has been specified, that is, $m = 1$, the validity of equation (47), representing the quintessence of this article, is verified numerically.

For the sake of completeness, also the eigenvalues of

$$[\boldsymbol{K} - \omega^2 \boldsymbol{M}]$$

are computed. Since this eigenproblem is characterized by the absence of constraint conditions, it is termed the corresponding "unconstrained problem". In the given case, the respective eigenvalues are obtained as

$$\omega_1^{*2} = 0{,}2475\,, \qquad \omega_2^{*2} = 1{,}0000\,, \qquad \omega_3^{*2} = 1{,}8859\,. \tag{60}$$

It is seen that

$$\omega_1^{*2} < \omega_1^2 < \omega_2^{*2} < \omega_2^2 < \omega_3^{*2} \tag{61}$$

as is required by the eigenvalue separation property [see equation (3)].

3.2 Buckling of a narrow plate strip subjected to dead load

Fig. 4 shows the finite element model of a narrow plate strip of length l, width b, and thickness h. The plate strip is assumed to be fixed at one end and free at the other one. The adjective "narrow" should indicate that application of beam theory to the present problem would yield an acceptable approximation of the results which were obtained by means of plate theory.

The discretized plate strip consists of 20 triangular finite elements of the same type as has been used for the solution of the first example. In order to satisfy the aforementioned C^1-continuity requirement at the interelement

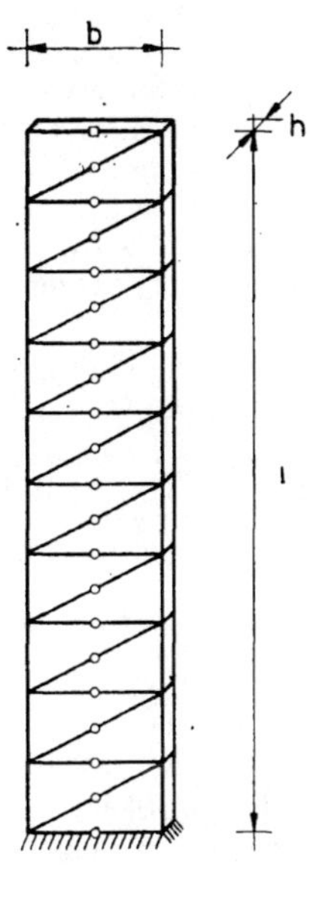

Fig. 4. Finite element model of a narrow plate strip

boundaries, constraint conditions analogous to the one discussed in the preceding subsection must be specified at the midside nodes. According to Fig. 4, the number of such conditions, m, is equal to 20. For the sake of completeness, an "associated constrained problem" will also be solved. It is characterized by specification of an additional constraint condition, namely, "zero-slope" at the midside node on the free edge of the plate strip (see Fig. 4). Consequently, for the "associated constrained problem" $m = 21$.

The present buckling analysis is restricted to consideration of dead load, g. Note that for instability analyses the matrix $\boldsymbol{H}$, originally defined in equation (35.1), is re-defined as follows:

$$\boldsymbol{H} = \boldsymbol{K} + \chi \boldsymbol{N}_1(\overline{\boldsymbol{q}}) \,. \tag{62}$$

In (62), $\boldsymbol{N}_1(\overline{\boldsymbol{q}})$ is the well-known first (linear) geometric stiffness matrix, evaluated for displacements $\overline{\boldsymbol{q}}$, resulting from a reference load; χ is a proportionality factor — the eigenvalue.

Table 1 contains the result of the iterative determination of a lower and an upper bound to the second eigenvalue, utilizing the modified Sturm-sequence check. Selecting a starting value $\overline{\chi}_0$, determining a lower bound to the second eigenvalue, $\overline{\chi}_1 = \overline{\chi}^{(2)}_{MIN}$, by choosing a multiple of $\overline{\chi}_0$ such that

$$k_1 = s_1 - m = 1 \tag{63}$$

Table 1. Buckling of narrow plate strip subjected to dead load — iterative determination of bounds to second eigenvalue utilizing the modified Sturm-sequence check

	Starting Value (user specified)	Multiples of Starting Value		Successive bisection of interval $\langle \overline{\chi}_{MIN}, \overline{\chi}_2 \rangle$		
	$\overline{\chi}_0$	$\overline{\chi}_1 = 10\overline{\chi}_0 = \overline{\chi}^{(2)}_{MIN}$	$\overline{\chi}_2 = 10\overline{\chi}_1$	$\overline{\chi}_3 = \frac{\overline{\chi}^{(2)}_{MIN} + \overline{\chi}_2}{2}$	$\overline{\chi}_4 = \frac{\overline{\chi}^{(2)}_{MIN} + \overline{\chi}_3}{2}$	$\overline{\chi}_5 = \frac{\overline{\chi}^{(2)}_{MIN} + \overline{\chi}_4}{2} = \overline{\chi}^{(2)}_{MAX}$
$\overline{\chi}_i$; $i = 0, 1, \ldots, 5$	5,0000	50,0000	500,0000	275,0000	162,5000	106,2500
$\mathrm{Det}(\overline{\chi}_i)$	0,2993	−0,1992	−0,6742	−0,2369	−0,3055	0,7392
s_i	20	21	25	23	23	22
$k_i = s_i - m$	0	1	5	3	3	2

Table 2. Buckling of a narrow plate strip subjected to dead load — comparison of eigenvalues of the original problem with eigenvalues of the associated constrained problem

	Original problem (Number of constraint conditions: $m = 20$)	Associated constrained problem (Number of constraint conditions: $m + 1 = 21$)	
Number of eigenvalue	$\chi_i = (gl)_i \frac{l^2}{EI}$	$\chi_i^* = (gl)_i^* \frac{l^2}{EI}$	Number of eigenvalue
1	7,8374*		
2	55,988	18,932	1
3	148,69	81,835	2
4	286,62	189,41	3

*) Analytical: 7,837 [see, for example, Ref. [4]]

and selecting a multiple of $\overline{\chi}^{(2)}_{MIN}$, $\overline{\chi}_2$, such that $k_2 > k_1$, permits computation of an upper bound to the desired eigenvalue, $\overline{\chi}^{(2)}_{MAX}$, by means of successive bisection of the interval $\langle \overline{\chi}^{(2)}_{MIN}, \overline{\chi}_2 \rangle$ until

$$k_i = s_i - m = 2 = k_1 + 1 \,. \tag{64}$$

For each value $\overline{\chi}_i$, $i = 0, 1, \ldots, 5$, Table 1 contains the value of the buckling determinant, Det $(\overline{\chi}_i)$, the number of negative terms on the main-diagonal of the triangularized buckling matrix, s_i, and the number of eigenvalues smaller than $\overline{\chi}_i$, k_i.

After determination of the interval $\langle \overline{\chi}^{(2)}_{MIN}, \overline{\chi}^{(2)}_{MAX} \rangle$, the search for the desired root of the characteristic polynomial, $\chi^{(2)}$, is continued. Details of the employed determinant-search procedure are given in [3].

Table 2 contains a comparison of the first four eigenvalues of the original problem, χ_i, $i = 1, 2, 3, 4$, with the first three eigenvalues of the previously described associated constrained problem, χ_j^*, $j = 1, 2, 3$. It is seen that

$$\chi_1 < \chi_1^* < \chi_2 < \chi_2^* < \chi_3 < \chi_3^* < \chi_4 \,, \tag{65}$$

as is required by the eigenvalue separation property [see equation (3)].

4. Summary and Conclusions

It was shown that the classical Sturm-sequence check can easily be adapted to eigenvalue problems resulting from variational principles with subsidiary conditions in the form of linear constraint conditions for the unknown degrees of freedom. All that needs to be done in order to determine the number of eigenvalues smaller than a shift to the right of the largest one of these eigenvalues is to subtract the number of constraint conditions, which are known a priori, from the number of negative terms on the main-diagonal of the triangularized characteristic matrix. Incorporation of the modified Sturm-sequence check into an existing computer program which is based on an analysis technique derived from a variational principle with subsidiary conditions of the aforementioned form, is trivial.

The results of the theoretical investigation were verified within the framework of a detailed numerical study.

References

1 BATHE, K.-J.; WILSON, E. L., Numerical Methods in Finite Element Analysis, Prentice-Hall, Englewood Cliffs, New Jersey 1976.
2 MANG, H. A.; GALLAGHER, R. H., A critical assessment of the simplified hybrid displacement method, Internat. J. numer. Meth. Engng. **11**, 145–167 (1977).
3 WALTER, H., Zur numerischen Bestimmung von Beuldicken dünner Platten und Schalen bei Berücksichtigung einer Eigenschaft der Sturm'schen Folgen, Diplomarbeit, Institut für Baustatik und Festigkeitslehre, Technische Universität Wien 1980.
4 TIMOSHENKO, S. P.; GERE, J. M., Theory of Elastic Stability, 2nd edition, McGraw-Hill, New York 1961.

Eingereicht: 5. 5. 1980

Anschriften: Ao. Univ. Prof. Dipl.-Ing. Dr. techn. et Ph. D. HERBERT MANG; Dipl.-Ing. HERBERT WALTER, Institut für Baustatik und Festigkeitslehre, Technische Universität Wien, Karlsplatz 13, A-1040 Wien, Österreich

ZAMM 61, 559 – 566 (1981)

K. KNOTHE / H. HERRMANN / W. MÜLLER

Numerische Stabilität von Plattenelementen auf der Grundlage hybrider Deformationsmodelle

Bei Plattenberechnungen mit Elementen auf der Grundlage der vereinfachten, hybriden Deformationsmethode, wobei auf die Erfüllung von C^1-Stetigkeitsforderungen verzichtet wird, können numerische Instabilitäten auftreten. Die Ursache dieser Instabilitäten wird untersucht und es werden Abhilfemaßnahmen aufgezeigt.

The analysis of plate bending problems using the so called simplified hybrid displacement method, whereby the C^1-continuity requirement must not be fulfilled, can result in numerical instabilities. The reasons of these instabilities are investigated and measures to avoid the instabilities are shown.

При рассчёте изгиба пластин так называемым упрощённым гибридным методом перемещений, пренебрегая требованиями C^1-непрерывности, результат может быть численно неустойчивым. Исследуются причины этих неустойчивостей и возможности их устранения.

Aufgrund der numerischen Untersuchungen, die u. a. von KIKUCHI und ANDO [1, 2] und von den Verfassern dieser Arbeit [3, 4, 5] veröffentlicht wurden, wurden sogenannte vereinfachte, hybride Deformationsmodelle als eine der Möglichkeiten angesehen, die Schwierigkeiten, die sich bei der Erfüllung der C^1-Stetigkeitsforderung bei der Entwicklung von finiten Elementen für schubstarre Platten und Schalen auf der Grundlage des Prinzips vom Minimum der potentiellen Energie ergeben, zu überwinden. In einer Veröffentlichung von MANG und GALLAGHER [6] wurde erstmals darauf hingewiesen, daß bei derartigen Verfahren numerische Instabilitäten auftreten können. In der zweiten Hälfte eines an der TU Berlin durchgeführten DFG-Forschungsvorhabens wurden die Ursachen der bei Elementen auf der Grundlage der vereinfachten, hybriden Deformationsmethode (VHDM) auftretenden Instabilitäten und mögliche Abhilfemaßnahmen untersucht und es wurden Schlußfolgerungen für die weitere Beschäftigung mit hybriden Verfahren gezogen. Über diese Untersuchungen wird im folgenden berichtet.

1. Begriffe, Definitionen, Funktional

Zum besseren Verständnis sollen einleitend die verwendeten Begriffe und das verwendete Funktional erläutert werden [5, 7, 8]. Bei Finite-Element-Verfahren auf der Basis des Prinzips vom Minimum der potentiellen Energie muß sichergestellt sein, daß die verwendeten Ansatzfunktionen die notwendigen Stetigkeitsforderungen — bei schubstarren Platten und Schalen sind es C^1-Stetigkeitsforderungen — erfüllen. Ist eine A-Priori-Erfüllung dieser Forderungen nur schwer oder unter Umständen gar nicht möglich, so kann man sie mit der LAGRANGEschen Multiplikatorentechnik in das Variationsprinzip einführen. Im Variationsprinzip treten dann die LAGRANGEschen Multiplikatoren, die nur auf den Interelementgrenzen (Gitterlinien, Randstreifen) definiert sind, als zusätzliche Zustandsgrößen auf. Für derartige Funktionale wird in Anlehnung an PIAN [9] der Begriff „hybride Funktionale" verwendet.

Das Finite-Element-Modell zu dem von uns untersuchten hybriden Funktional wird durch Abb. 1 illustriert. Bei diesem Funktional werden die folgenden unabhängigen Zustandsgrößen eingeführt:

$\boldsymbol{u}$ ist der Verschiebungszustand im Bereich B_n des Elementes,

$\boldsymbol{u}_L$ ist ein zusätzlicher Verschiebungszustand, der auf die Gitterlinien beschränkt ist. Bei schubstarren Platten besteht dieser Verschiebungszustand aus der Querverschiebung w und der Normalneigung $\partial w/\partial n$. Er ist in Abbildung 1 durch einen Streifen gekennzeichnet.

Den Schnittmomentenzustand $\boldsymbol{\sigma}$ zum Verschiebungszustand $\boldsymbol{u}$ erhält man, indem man mit Hilfe eines Differentialoperators $\boldsymbol{D}_{\varepsilon u}$ zuerst die Verzerrungen (Krümmungen) und anschließend durch Multiplikation mit einer Elastizitätsmatrix $\boldsymbol{E}$ die Schnittmomente (Biege- und Drillmomente) ermittelt. Die Randwerte des Verschiebungs- und Schnittmomentenzustandes werden mit dem Index r gekennzeichnet. Im Funktional werden weiterhin für die in einem Bereich B_n wirkenden Flächenlasten die Bezeichnung $\overline{\boldsymbol{p}}$ und für die auf einem belasteten Außenrand R_σ wirkenden Randbelastungen die Bezeichnung $\overline{\boldsymbol{\sigma}}_R$ eingeführt. Das Funktional, das erstmals von TONG [10] eingeführt wurde, lautet dann:

$$\Pi_{mD3} = \sum_n \int_{B_n} \left[\tfrac{1}{2} (\boldsymbol{D}_{\varepsilon u}\boldsymbol{u})^T \boldsymbol{E}(\boldsymbol{D}_{\varepsilon u}\boldsymbol{u}) - \boldsymbol{u}^T\overline{\boldsymbol{p}}\right] \mathrm{d}F + \sum_n \oint_{R_n} (\boldsymbol{u}_L - \boldsymbol{u}_r)^T (\boldsymbol{E}\boldsymbol{D}_{\varepsilon u}\boldsymbol{u})_r \,\mathrm{d}s - \oint_{R\sigma} \boldsymbol{u}_L^T\overline{\boldsymbol{\sigma}}_R \,\mathrm{d}s = \text{stat}\,.$$

Während der erste und der dritte Term bereits beim Prinzip vom Minimum der potentiellen Energie auftreten, stellt Term 2 einen Zusatzterm dar, der verschwindet, wenn die Randwerte des Bereichsverschiebungszustandes mit dem Gitterlinienverschiebungszustand übereinstimmen.

Bei dem Funktional müssen zwei Nebenbedingungen eingehalten werden:

(1) Der Gitterlinienverschiebungszustand muß auf dem Randabschnitt R_u, auf dem Verschiebungen vorgegeben sind, die dort vorgeschriebenen Werte annehmen:

$$\boldsymbol{u}_L - \boldsymbol{u}_R \doteq 0 \quad \text{auf} \quad R_u\,.$$

(2) Der Verschiebungszustand $\boldsymbol{u}$ in einem Element darf keine vom Gitterlinienverschiebungszustand $\boldsymbol{u}_L$ unabhängigen Starrkörperverschiebungen ausführen.

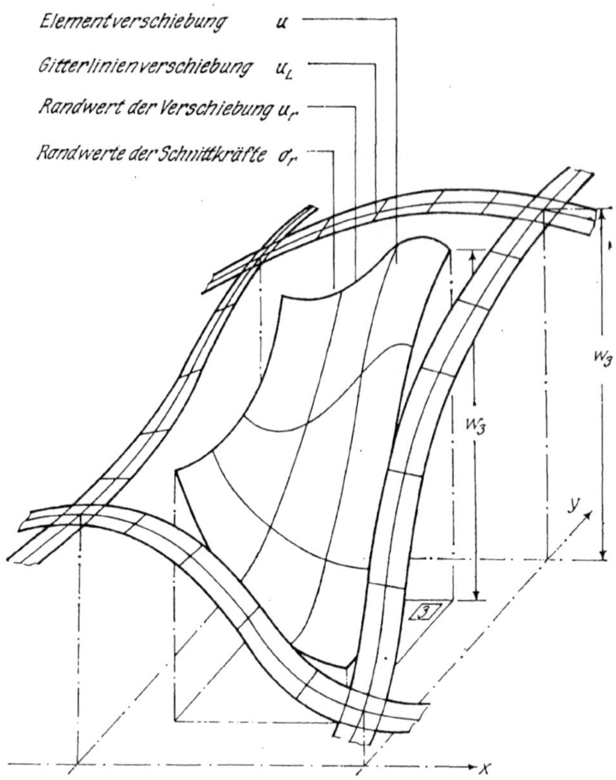

Abb. 1. Hybrides Deformationsmodell

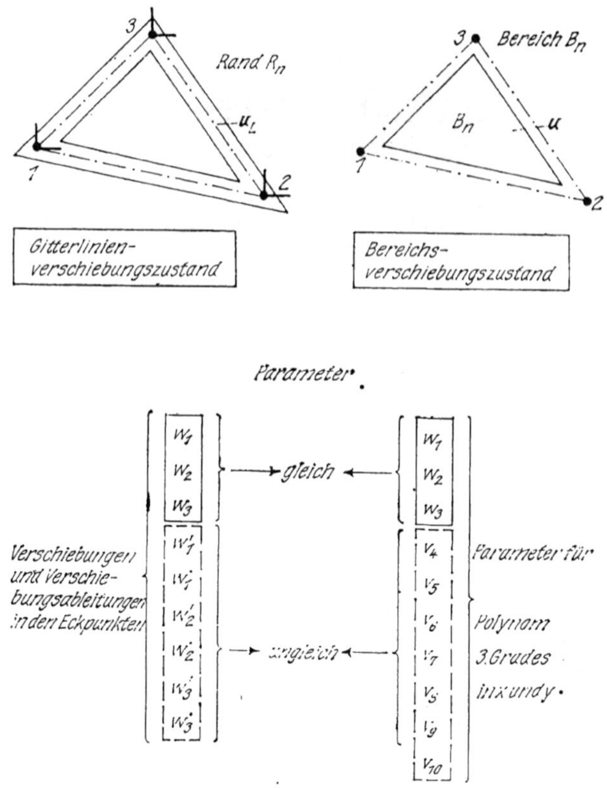

Abb. 2. Parameterwahl beim Plattendreieckselement der normalen, hybriden Deformationsmethode (NHDM)

2. Wahl der Ansatzfunktionen

Die zweite Forderung läßt sich, wie man am Beispiel eines Dreieckselementes (Abb. 2) sieht, leicht bei der Wahl der Ansatzfunktionen erfüllen. Für den Gitterlinien- und den Bereichsverschiebungszustand werden die Eckpunktverschiebungen gleichermaßen als Parameter gewählt. Diese Vorgehensweise wollen wir als normale, hybride Deformationsmethode (NHDM) bezeichnen.

Bei der sogenannten vereinfachten, hybriden Deformationsmethode (VHDM) werden auch noch die Verschiebungsableitungen in den Eckpunkten zugleich als Parameter für den Gitterlinien- und für den Bereichsverschiebungszustand verwendet. Die Parameter v_4 bis v_9 werden dabei durch die Parameter w_1' bis w_3' ersetzt. Das hat zur Folge, daß der Bereichs- und der Gitterlinienverschiebungszustand und damit auch die Verschiebungszustände benachbarter Elemente in den Knotenpunkten C^1-stetig ineinander übergehen, während entlang der Seiten zwischen den Knotenpunkten weiterhin Inkompatibilitäten auftreten können. Die VHDM besitzt zwei entscheidende Vorteile:

(1) Man erhält nach Zwischenelimination der inneren Unbekannten v_{10} Steifigkeitsmatrizen, die sich ohne weiteres in übliche Programmsysteme der Methode der Finiten Elemente einbauen lassen.

(2) Bei Dreieck- und Rechteckselementen lassen sich die Steifigkeitsmatrizen in geschlossener Form angeben, wodurch man erhebliche Rechenzeit beim Aufstellen des Gleichungssystems spart.

Numerische Untersuchungen zur VHDM wurden von Kikuchi und Ando [1, 2] und an der TU Berlin [3, 4, 5] durchgeführt. Bei den Standardfällen (Quadratplatte unter Flächenlast und Einzellast) ergab sich bei charakteristischen Verschiebungen eine Fehlerordnung von 2. Das Plattendreieckselement der VHDM erwies sich in diesen Fällen hinsichtlich der erreichbaren Genauigkeit allen anderen Plattenelementen zumindest gleichwertig, wenn als Vergleichskriterium der erforderliche Rechenaufwand herangezogen wurde [4]. Die gleiche Tendenz stellte sich bei der Verschiebungsberechnung von Parallelogramm-, Kreis-, und Kreisringplatten ein.

3. Instabilitäten

Erstmals wurde von Mang und Gallagher [6] über numerische Instabilitäten bei der VHDM berichtet. Sobald im Elementverband Dreieckselemente auftreten, die merklich von der Gleichseitigkeit oder von der Gleichschenklichkeit bei rechtwinkligen Dreiecken abweichen, kann die positive Definitheit der Systemsteifigkeitsmatrix verlorengehen. Haugeneder machte in einer von Mang angeregten Dissertation [11] dafür das Ausgangsfunktional, Gleichung (1), verantwortlich. Das Funktional sei unbrauchbar, da der Lagrangesche Multiplikator $\boldsymbol{u}_L$ eine Verschiebungsgröße und damit nicht mehr dual zu der beim Prinzip vom Minimum der potentiellen Energie verletzten geometrischen Übergangsbedingungen ist. Zu beantworten war damit u. a. die Frage:

— Ist, wie von HAUGENEDER behauptet, das Ausgangsfunktional unbrauchbar, oder
— ist die Versagensursache „nur" der Übergang von der NHDM zur VHDM?

Die Untersuchung der Ursachen der aufgetretenen Instabilitäten wird der Einfachheit halber an einem Rechteckselement durchgeführt, bei dem das Seitenverhältnis als einziger geometrischer Parameter auftritt. Um überhaupt zu einem Element der hybriden Deformationsmethode zu gelangen, soll der Verschiebungszustand im Bereich B_n durch Ansatzfunktionen approximiert werden, die aus einem Polynom vom Grad 4 in x und y aufgebaut werden. Mit derartigen Ansatzfunktionen läßt sich, anders als mit bikubischen, HERMITEschen Interpolationspolynomen, keine C^1-Stetigkeit erreichen. Als Parameter sind neben den Verschiebungen und Verschiebungsableitungen in den vier Eckpunkten (zwölf Eckpunktparameter) drei zusätzliche, innere Unbekannte erforderlich. Der Gitterlinienverschiebungszustand wird nur durch die zwölf Eckpunktparameter beschrieben. Damit läßt sich die Verschiebung w_L kubisch, die Normalableitung $(\partial w/\partial n)_L$ linear approximieren (vgl. auch Abb. 3).

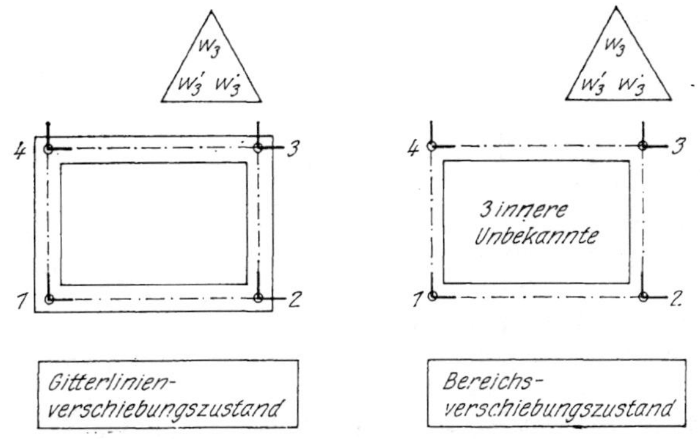

Abb. 3. Parameterwahl beim Plattenrechteckselement der vereinfachten, hybriden Deformationsmethode (VHDM)

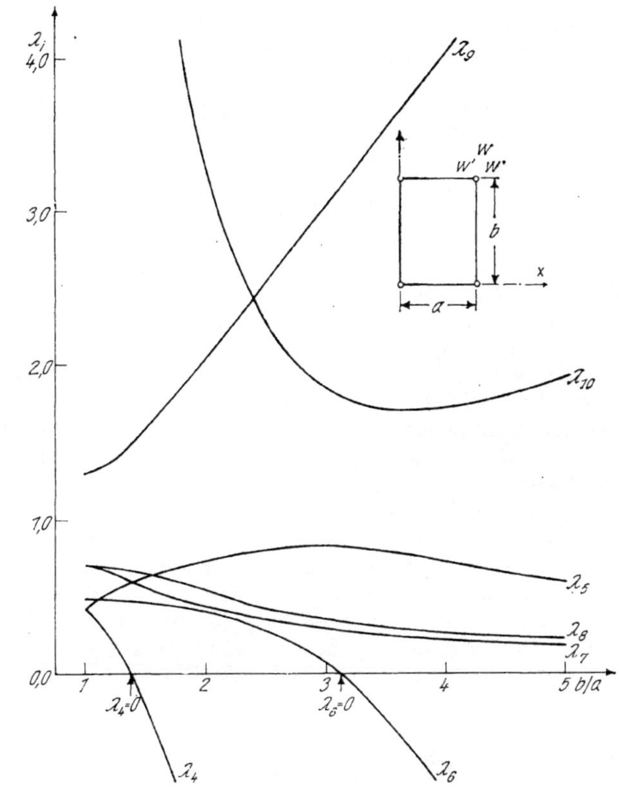

Abb. 4. Eigenwertspektrum des Plattenrechteckselementes der VHDM für $b/a > 0$ ($\lambda_1 = \lambda_2 = \lambda_3 = 0$)

Zur Untersuchung der numerischen Stabilität dieses Rechteckselementes wurden nach Elimination der drei inneren Unbekannten durch Lösen der Eigenwertaufgabe

$$[\boldsymbol{S} - \lambda \boldsymbol{I}]\,\boldsymbol{y} = \boldsymbol{0}$$

die Eigenwerte λ_i und die Eigenvektoren $\boldsymbol{\varphi}_i$ der Elementssteifigkeitsmatrix $\boldsymbol{S}$ ermittelt. Der Elementverschiebungsvektor $\boldsymbol{y}$, der nur noch Knotenverschiebungen und Verschiebungsableitungen enthält, kann jetzt mit der Modalmatrix $\boldsymbol{\Phi}$, in der alle Eigenvektoren $\boldsymbol{\varphi}_i$ zusammengefaßt sind, geschrieben werden als

$$\boldsymbol{y} = \boldsymbol{\Phi} \boldsymbol{a} .$$

Für die generalisierte Formänderungsenergie eines Elementes, in der sowohl die Bereichs- als auch die Randanteile erfaßt werden, erhält man nach einer entsprechenden Transformation

$$2\Pi_{i,\,\mathrm{gen}}^{(\mathrm{Element})} = \boldsymbol{y}^T \boldsymbol{S} \boldsymbol{y} = \boldsymbol{a}^T \boldsymbol{\Phi}^T \boldsymbol{S} \boldsymbol{\Phi} \boldsymbol{a} = \boldsymbol{a}^T \boldsymbol{\Lambda} \boldsymbol{a} ,$$

wobei $\boldsymbol{\Lambda}$ eine nur diagonal mit den Eigenwerten λ_i besetzte Matrix ist. Man erkennt hieran unmittelbar, daß die positive Definitheit der Systemsteifigkeitsmatrix verlorengeht, sobald außer den drei Null-Eigenwerten, die zu den Starrkörpereigenformen gehören, weitere Eigenwerte erst verschwinden und dann negativ werden.

In Abb. 4 ist das Eigenwertspektrum für die Steifigkeitsmatrix des Rechtseckselementes der VHDM mit zwölf Freiheitsgraden in Abhängigkeit vom Seitenverhältnis (für $b/a > 0$) angegeben. Nicht eingetragen sind die verschwindenden Eigenwerte $\lambda_1, \lambda_2, \lambda_3 \equiv 0$.

Bei einem Seitenverhältnis $b/a = 1{,}372$ wird ein erster Eigenwert, bei einem Seitenverhältnis $b/a = 3{,}1$ ein weiterer Eigenwert negativ. Zwei weitere Eigenwerte werden negativ, wenn man b/a durch a/b ersetzt. Ein einzelnes

Element ist damit bereits von einem Seitenverhältnis 1,372 an unbrauchbar. Es ist zwar zu erwarten, daß im Elementverband eine gewisse Stabilisierung erreicht wird, so daß das kritische Seitenverhältnis etwas hinausgeschoben wird, aus Sicherheitsgründen wird man aber fordern, daß ein Seitenverhältnis von 1,372 nicht überschritten wird. Da derartige Seitenverhältnisse in der Praxis durchaus auftreten, ist das Element unbrauchbar.

Ähnlich liegen die Verhältnisse beim Plattendreieckselement zur VHDM [6].

4. Versuch einer Erklärung der aufgetretenen Instabilitäten

In Abb. 5 sind die beiden Eigenformen, bei denen die zugehörigen Eigenwerte negativ werden, in der Nähe des jeweils kritischen Seitenverhältnisses dargestellt. Im folgenden soll für die Eigenform zu $\lambda_4 = 0$ die generalisierte Formänderungsenergie näher untersucht werden. Als generalisierte Formänderungsenergie wird der auf ein Element i bezogene Anteil des Funktionals von Gleichung (1) unter Einbezug der Randterme betrachtet:

$$\Pi_{i,\,\mathrm{gen}}^{(\mathrm{Element})} = \int_{B_n} \tfrac{1}{2} (\boldsymbol{D}_{\varepsilon u}\boldsymbol{u})^T \boldsymbol{E}(\boldsymbol{D}_{\varepsilon u}\boldsymbol{u})\,\mathrm{d}F_n + \oint_{R_n} (\boldsymbol{u}_L - \boldsymbol{u}_r)^T (\boldsymbol{E}\boldsymbol{D}_{\varepsilon u}\boldsymbol{u})_r\,\mathrm{d}s_n\,.$$

Das Umlaufintegral wird nun noch skalar formuliert, wobei mit $\bar{q}_n$ die Ersatzquerkraft und mit m_n das Normalmoment am Rand des Bereiches bezeichnet wird:

$$\Pi_{i,\,\mathrm{gen}}^{(\mathrm{Element})} = \int_{B_n} \frac{1}{2} (\boldsymbol{D}_{\varepsilon u}\boldsymbol{u})^T \boldsymbol{E}(\boldsymbol{D}_{\varepsilon u}\boldsymbol{u})\,\mathrm{d}F_n + \oint_{R_n} (w_L - w_r)\,\bar{q}_n\,\mathrm{d}s_n + \oint_{R_n} \left[\left(\frac{\partial w}{\partial n}\right)_r - \left(\frac{\partial w}{\partial n}\right)_L\right] m_n\,\mathrm{d}s_n\,.$$

In der klassischen Deformationsmethode, d. h. bei einer Formulierung auf der Grundlage des Prinzips vom Minimum der potentiellen Energie, tritt nur der erste Term auf, der stets positiv ist, während der zweite und dritte Term verschwinden, da wegen der C^1-Stetigkeit der Ansatzfunktionen Verschiebungs- und Neigungsdifferenzen nicht auftreten. Bei der VHDM verschwindet aufgrund der speziellen Wahl der Ansatzfunktionen zwar der zweite Term, der dritte Term wird bei der kritischen Eigenform aber negativ und wirkt sich damit abmindernd auf die stets positive Formänderungsenergie des ersten Terms aus.

Anhand von Abb. 6 läßt sich erläutern, wie diese Abminderung zustande kommt. Einen Beitrag zu dem Umlaufintegral erhält man nur aus den Rändern $x = \mathrm{const.}$ Der Randwert der Normalneigung verläuft hier parabolisch, während die Gitterliniennormalneigung, die mit den Randwerten der Normalneigung in den Eckpunkten übereinstimmt, linear verläuft. Zwischen beiden Verläufen verbleibt eine Differenz, die auch im Mittel nicht verschwindet. Das Normalmoment m_n ist konstant und besitzt ein solches Vorzeichen, daß der Term insgesamt negativ wird und damit abmindernd wirkt.

Eine der Ursachen des Versagens ist mithin, daß der Bereichsverschiebungszustand so gewählt wurde, daß er mit dem Gitterlinienverschiebungszustand in den Knotenpunkten (Elementeckpunkten) auch hinsichtlich der Verschiebungsableitungen übereinstimmt. Bei der kritischen Eigenform gelingt es damit nicht, die Differenz aus dem Randwert der Normalneigung und der Gitterlinienneigung wenigstens im Mittel zum Verschwinden zu bringen. Generell kann man feststellen, daß der gesamte Verschiebungszustand eines Elementes offensichtlich durch zu wenige Freiheitsgrade beschrieben wird, um alle EULER-Gleichungen eines Elementes (Gleichgewichtsbedingung, geometrische Randbedingung, statische Randbedingungen), im Integralmittel zu erfüllen.

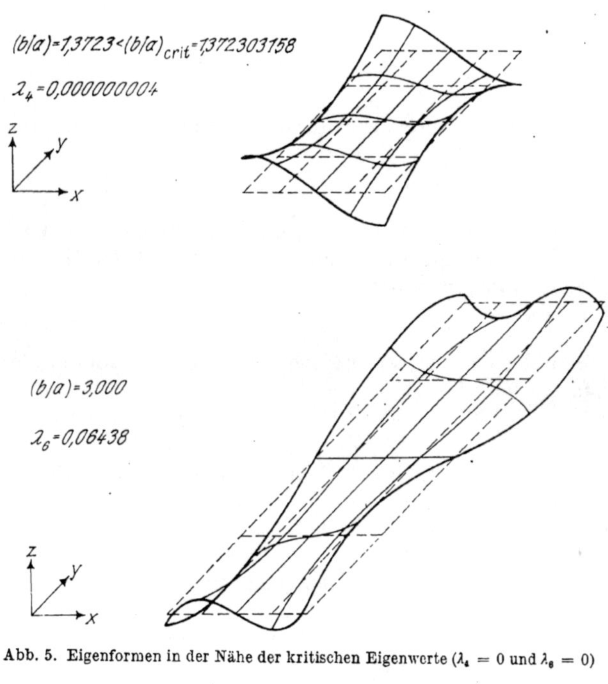

Abb. 5. Eigenformen in der Nähe der kritischen Eigenwerte ($\lambda_4 = 0$ und $\lambda_6 = 0$)

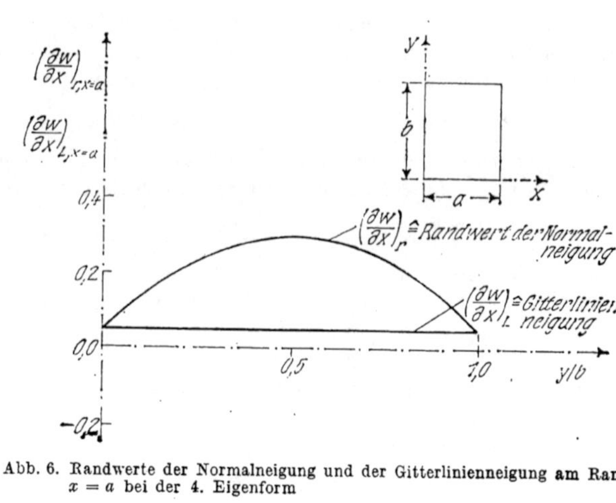

Abb. 6. Randwerte der Normalneigung und der Gitterlinienneigung am Rand $x = a$ bei der 4. Eigenform

5. Abhilfemaßnahmen

Es ist daher zu vermuten, daß der Instabilitätseffekt verschwindet, wenn man die Bereichs- und den Gitterlinienverschiebungszustand so wählt, daß sie in den Knotenpunkten nur noch in den Verschiebungen, nicht hingegen in den Verschiebungsableitungen übereinstimmen. Die Parameter zur Beschreibung des Gitterlinien-Verschiebungszustandes, Verschiebungen und Verschiebungsableitungen in den vier Knotenpunkten, werden in einem Vektor $\boldsymbol{w}$ zusammengefaßt:

$$\boldsymbol{w}^T = \{w_1, w_1', w_1^{\cdot}; w_2, w_2', w_2^{\cdot}; w_3, w_3', w_3^{\cdot}; w_4, w_4', w_4^{\cdot}\} .$$

Der Bereichsverschiebungszustand soll weiterhin durch ein Polynom 4. Grades in x und y (15 Parameter) beschrieben werden. Außer den Eckpunktverschiebungen w_i sind hierzu elf innere Unbekannte erforderlich, die in einem Vektor $\boldsymbol{v}$ zusammengefaßt werden. Die generalisierte Formänderungsenergie für dieses Element der NHDM wird dann durch die folgende Beziehung beschrieben:

$$\Pi_{i,\,\text{gen}}^{(\text{Element})} = \frac{1}{2}\,\{\boldsymbol{w}^T, \boldsymbol{v}^T\}\begin{bmatrix}\boldsymbol{S}_w & \boldsymbol{S}_{wv}\\ \boldsymbol{S}_{wv}^T & \boldsymbol{S}_v\end{bmatrix}\begin{Bmatrix}\boldsymbol{w}\\ \boldsymbol{v}\end{Bmatrix}.$$

Die elf im Vektor $\boldsymbol{v}$ zusammengefaßten inneren Unbekannten wird man zweckmäßigerweise auf Elementebene zwischeneliminieren, was nur noch numerisch möglich ist. Für die generalisierte Formänderungsenergie erhält man dann den Ausdruck

$$\Pi_{i,\,\text{gen}}^{(\text{Element})} = \tfrac{1}{2}\,\boldsymbol{w}^T[\boldsymbol{S}_w - \boldsymbol{S}_{wv}\boldsymbol{S}_v^{-1}\boldsymbol{S}_{wv}^T]\,\boldsymbol{w} .$$

Für das neue Element wurde nach Zwischenelimination der elf inneren Unbekannten wiederum eine Eigenwertanalyse durchgeführt. Das Eigenwertspektrum ist in Abb. 7 wiedergegeben. Gravierende Abweichungen treten nur gegenüber den kritischen Eigenwertkurven der VHDM auf, von denen zwei in Abb. 7 eingetragen wurden. Man erkennt:

(1) Bei dem Element der NHDM treten keine Instabilitäten mehr auf, da alle Eigenwerte positiv bleiben.

(2) Gegenüber den bisher kritischen Eigenwerten der VHDM ist ein Ansteigen zu beobachten, was darauf hinweist, das das neue Element der NHDM steifer ist als das alte Element der VHDM.

Diese größere Steifigkeit des Elementes, die ein Instabilwerden nicht zuläßt, wirkt sich nun aber leider negativ auf die Genauigkeit aus. In der folgenden Abb. 8 sind für vier unterschiedliche Quadratplatten die Fehlerbeträge der Mittendurchsenkung über der Quadrantunterteilung angegeben. Zusätzlich dargestellt sind die Werte, die man mit

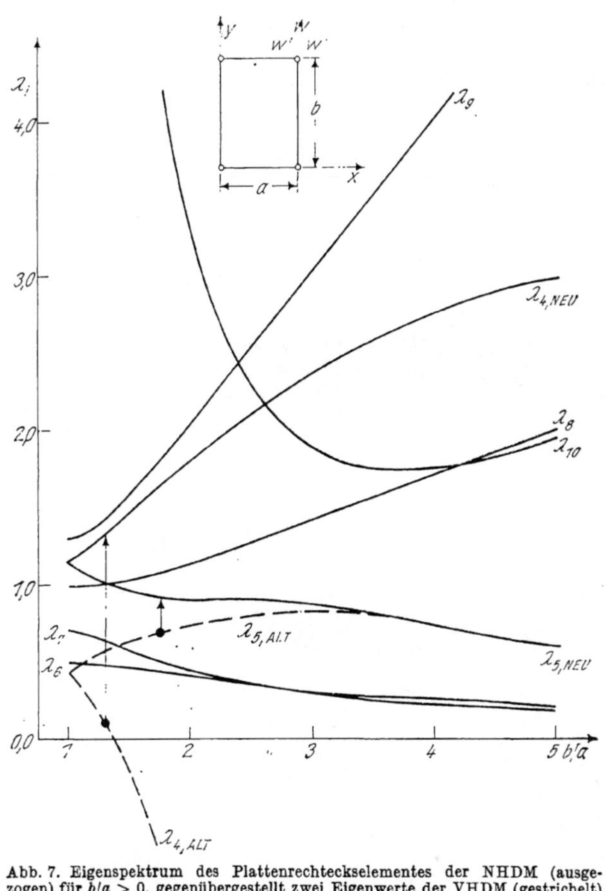

Abb. 7. Eigenspektrum des Plattenrechteckselementes der NHDM (ausgezogen) für $b/a > 0$, gegenübergestellt zwei Eigenwerte der VHDM (gestrichelt) ($\lambda_1 = \lambda_2 = \lambda_3 = 0$)

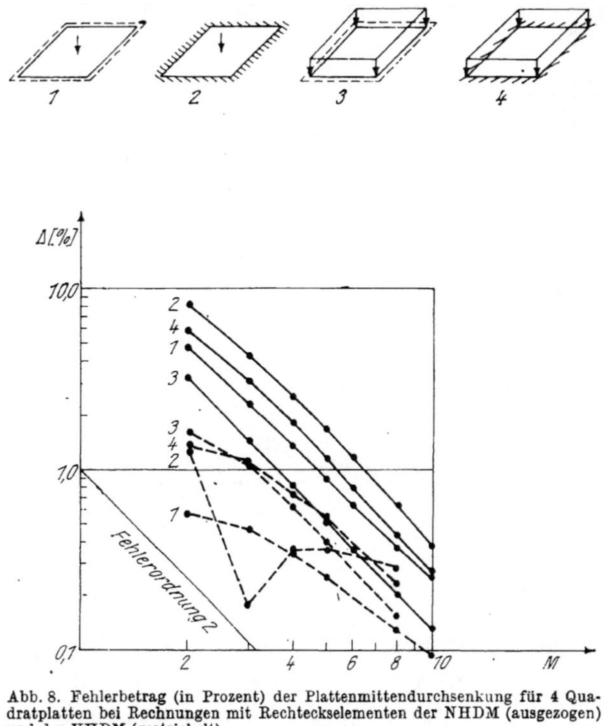

Abb. 8. Fehlerbetrag (in Prozent) der Plattenmittendurchsenkung für 4 Quadratplatten bei Rechnungen mit Rechteckselementen der NHDM (ausgezogen) und der VHDM (gestrichelt)

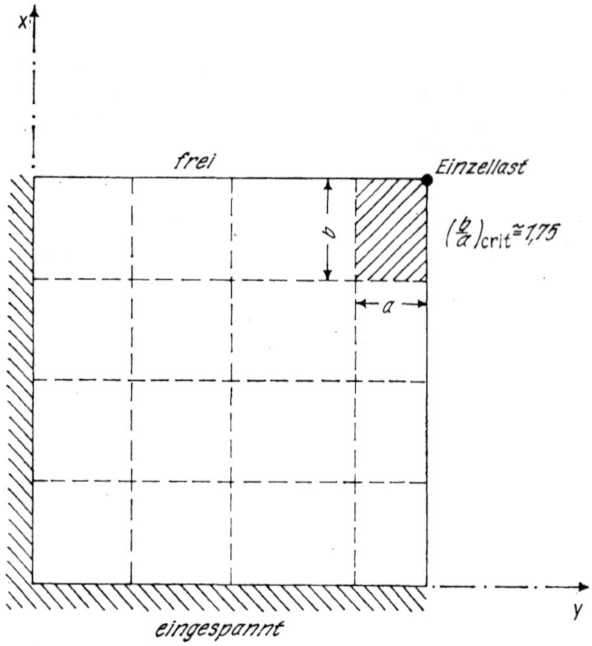

Abb. 9. Über Eck eingespannte Quadratplatte unter Einzellast (System, Elementunterteilung)

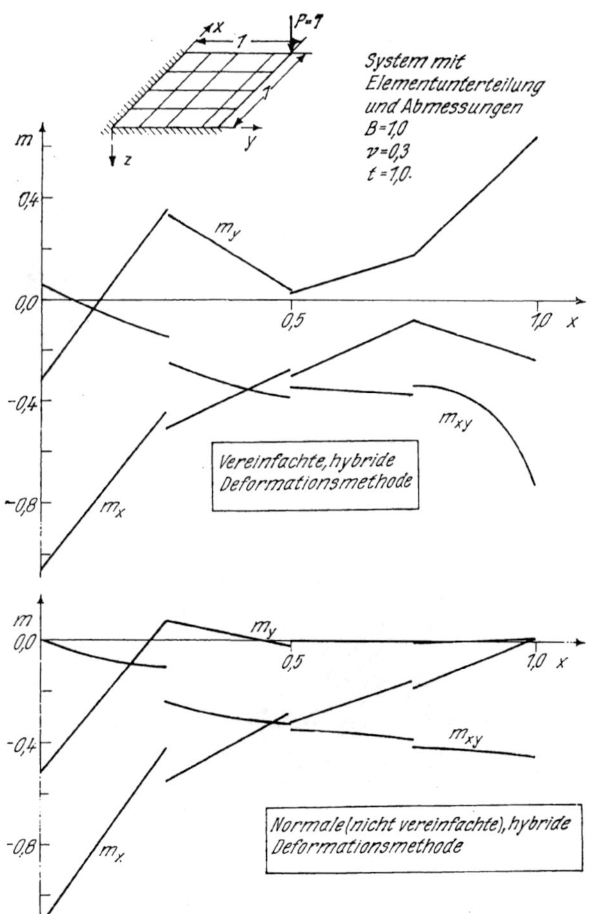

Abb. 10a. Biege- und Drillmomentenverläufe bei der über Eck eingespannten Quadratplatte unter Einzellast am Rand $y = 1$ für $b/a = 1{,}667 < (b/a)_{crit}$

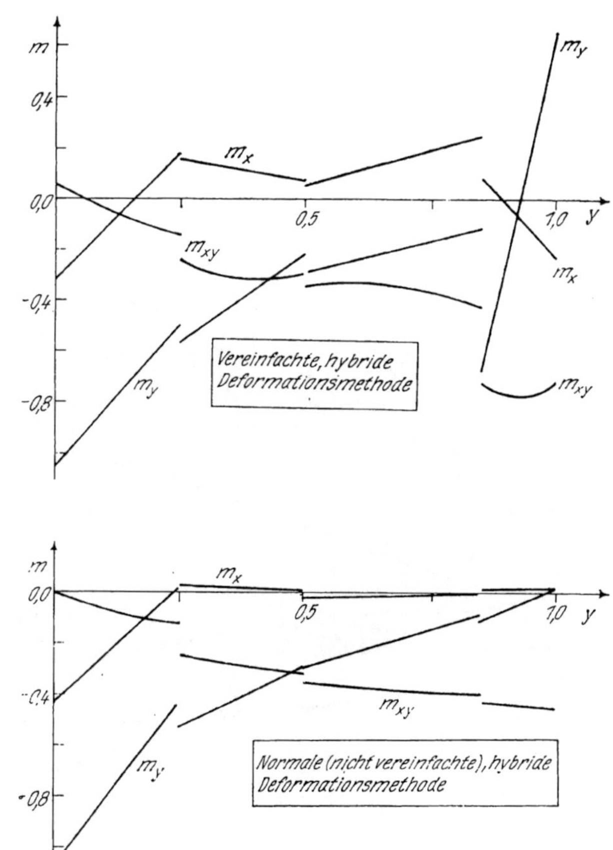

Abb. 10b. Biege- und Drillmomentenverläufe bei der über Eck eingespannten Quadratplatte unter Einzellast am Rand $x = 1$ für $b/a = 1{,}667 < (b/a)_{crit}$

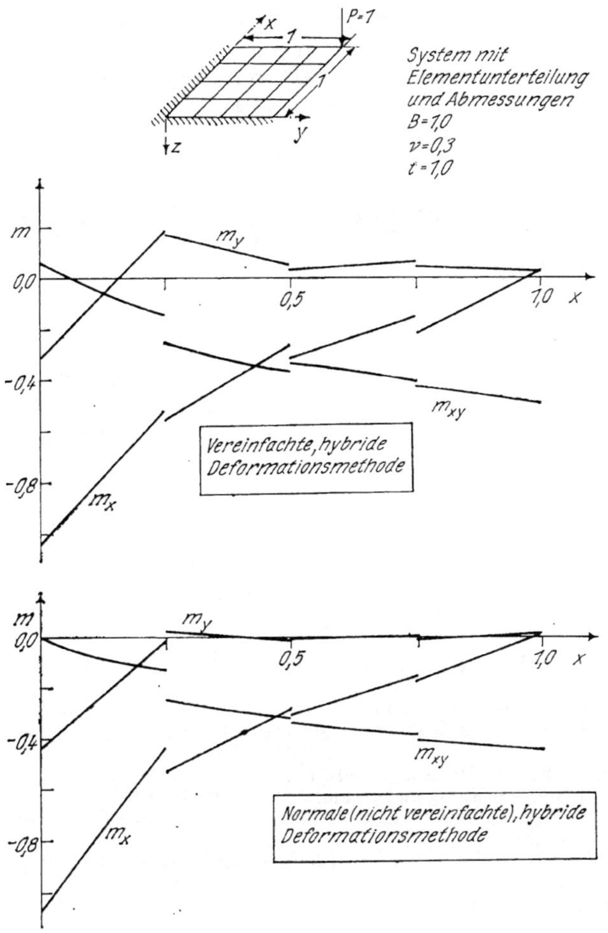

Abb. 11. Biege -und Drillmomentenverläufe bei der über Eck eingespannten Quadratplatte unter Einzellast am Rand $y = 1$ für $b/a = 1{,}0$

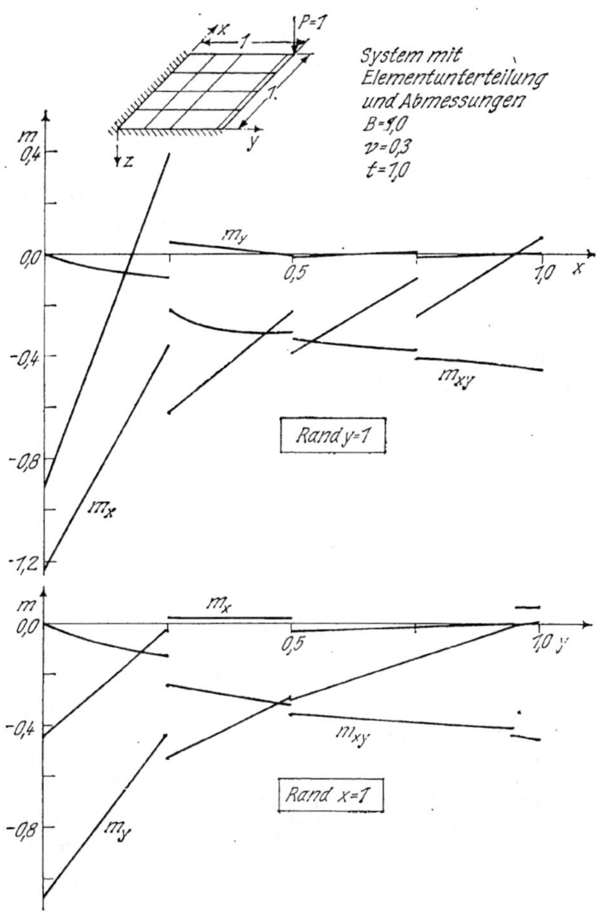

Abb. 12. Biege- und Drillmomentenverläufe bei der über Eck eingespannten Quadratplatte unter Einzellast für $b/a = 5{,}0$; Rechnung mit NHDM

dem Rechteckselement der VHDM erhält. Bei dem neuen Element stellt sich zwar sehr schnell die Fehlerordnung 2 ein, der Fehler ist aber betragsmäßig erheblich größer als bei dem Element der VHDM.

Recht eindrucksvoll ist eine Gegenüberstellung der Schnittmomentenverläufe, die sich bei der Berechnung mit beiden Elementen ergeben. Die Momente m_x, m_y und m_{xy} werden jeweils durch Differentiation aus dem Verschiebungszustand im Bereich ermittelt. Sie verlaufen dadurch natürlich, wie bei allen Elementen der Deformationsmethode, an den Interelementgrenzen unstetig. Als Beispiel wurde eine über Eck eingespannte, in der freien Ecke mit einer Einzellast beanspruchte Quadratplatte betrachtet (Abb. 9). Um die Auswirkung von nichtquadratischen Elementen überprüfen zu können, wurde eine Gitterlinie verschoben. Die Rechnung mit dem Element der VHDM wurde instabil, wenn das schmalste Element ein Seitenverhältnis von etwa $b/a = 1{,}75$ annahm.

In Abb. 10 sind für ein knapp darunterliegendes Seitenverhältnis ($b/a = 1{,}667$) die Momentenverläufe an einem freien Rand bei Berechnung mit Elementen der VHDM und der NHDM gegenübergestellt. Obwohl bei der VDHM die Systemsteifigkeitsmatrix noch positiv definit ist und übrigens auch die (hier nicht angegebenen) Verschiebungsverläufe noch sinnvoll erscheinen, erhält man unbrauchbare Schnittmomentenverläufe. Insbesondere wird die statische Randbedingung $m_y(x, 1) = 0$ nicht eingehalten. Das System mit Elementen der VHDM wird offensichtlich mit der Singularität an der Ecke, an der ein freier und ein eingespannter Rand zusammenstoßen, nicht fertig. Demgegenüber sind die Schnittmomentenverläufe, die man mit Elementen der NHDM erhält, brauchbar, auch wenn sie, wie bei allen Elementen der Deformationsmethode, an den Elementgrenzen Sprünge besitzen. Die statischen Randbedingungen werden außer in unmittelbarer Nähe der Singularität, recht gut eingehalten.

Daß die Schnittmomentenverläufe bei der VHDM zumindest hinsichtlich der Einhaltung der statischen Randbedingung unbrauchbarer sind, als bei der NHDM, erkennt man bereits an den Ergebnissen der Berechnungen mit Quadratelementen (Abb. 11). Selbst bei einem recht ungünstigen Seitenverhältnis von $b/a = 5$, bei dem die VDHM längst versagt, liefert die NHDM von der Tendenz her noch sinnvolle Ergebnisse, wobei nur die Unstetigkeiten an den Elementgrenzen stärker ausgeprägt sind (Abb. 12).

6. Zusammenfassung

Verantwortlich für die numerischen Instabilitäten der VHDM ist nicht, wie von Haugeneder behauptet, das Funktional, sondern die Tatsache, daß neben den Knotenverschiebungen auch die Knotenneigungen als gemeinsame Parameter für die Verschiebungen im Bereich und auf den Interelementgrenzen verwendet werden.

Verzichtet man auf diese Vereinfachung, so bleibt die Elementsteifigkeitsmatrix stets positiv definit, da das Element insgesamt steifer wird. Durch diese Erhöhung der Steifigkeit wird auch der Fehlerbetrag bei der Verschiebungsermittlung erheblich größer. Die Schnittmomentenverläufe, die bei der VDHM bereits vor dem Instabilwerden unbrauchbar werden, vergleiche auch [11], werden hingegen bei der NHDM, d. h. bei Verzicht auf die Vereinfachungen, erheblich besser.

Die Berechnungen wurden zwar nur für das Rechteckselement durchgeführt, es ist aber zu erwarten, daß sich beim Dreieckselement ähnliche Ergebnisse einstellen.

Der entscheidende Nachteil des Elementes der NHDM besteht darin, daß zur Ermittlung einer Elementsteifigkeitsmatrix, die sich in einem üblichen FEM-Programm verwenden läßt, auf Elementebene eine numerische Zwischenelimination durchgeführt werden muß. Aufgrund dieser Zwischenelimination ist es weder beim Rechtecksnoch beim Dreieckselement möglich, geschlossene Beziehungen für die Koeffizienten der Elementsteifigkeitsmatrix anzugeben. Der Rechenaufwand zum Aufstellen des Gleichungssystems steigt immens an, der erhebliche Vorteil des Elementes der VHDM ist damit verlorengegangen.

Wenn man sowieso gezwungen ist, den Zustand im Element durch eigene Parameter zu erfassen, die auf Elementebene zwischeneliminiert werden, dann empfiehlt es sich, gleich zu der von PIAN [12] vorgeschlagenen hybriden Gleichgewichtsmethode überzugehen, bei der der Zustand im Element durch Schnittkräfte, die die Gleichgewichtsbedingung befriedigen, beschrieben wird. Bei nichtlinearen Problemen wird man hingegen mit einem gemischten, hybriden Verfahren arbeiten [13].

Literatur

1 KIKUCHI, F.; ANDO, Y., A New Variational Functional for the Finite Element Method and its Application to Plate and Shell Problems, Nuclear Engineering Design **21**, 95–113 (1972).
2 KIKUCHI, F.; ANDO, Y., Some Finite Element Solutions for Plate Bending Problems by Simplified Hybrid Displacement Method, Nuclear Engineering Design **23**, 155–178 (1972).
3 KNOTHE, K.; HERRMANN, H., Theoretische Vorüberlegungen und numerische Untersuchungen zu einem Platten-Dreieckselement auf der Grundlage eines modifizierten DIRICHLETschen Variationsprinzips, ILR-Mitt. **24** (1976), Institut für Luft- und Raumfahrt, TU Berlin, 1976.
4 KNOTHE, K.; HERRMANN, H., Numerische Untersuchungen zu einem Platten-Dreieckselement auf der Grundlage eines modifizierten DIRICHLETschen Variationsprinzips, ILR-Mitt. 42 (1977), Institut für Luft- und Raumfahrt, TU Berlin 1977.
5 KNOTHE, K.; HERRMANN, H., Numerische Untersuchungen zu zwei Platten-Dreiecks-Elementen auf der Basis von Variationsprinzipien mit gelockerten Stetigkeitsforderungen, ZAMM **58**, S. 409–411 (1978).
6 MANG, H. A.; GALLAGHER, R. H., A Critical Assessment of Simplified Hybrid Displacement Method, Int. J. Num. Meth. Eng. **11**, S. 145–167 (1977).
7 KNOTHE, K., Ein modifiziertes DIRICHLETsches Variationsprinzip mit gelockerten Stetigkeitsforderungen für Platten, ZAMM **55** (1975), S. T 96 bis T 99.
8 KNOTHE, K., Ein Beitrag zur Systematisierung von Variationsprinzipien der Elastizitätstheorie und zu ihrer Verwendbarkeit für Finite-Element-Verfahren, ZAMM **58**, 405–412 (1978).
9 PIAN, T. H. H., How did I think of the Assumed Stress Hybrid Method, in: WASHIZU, K., Variational Methods in Elasticity (in Japanese), pp. 101–103 (1975).
10 TONG, P., New Displacement Hybrid Finite Element Modells for Solid Continua, Int. J. Num. Methods Eng. **2**, S. 73–83 (1970).
11 HAUGENEDER, E., Über Variationsprinzipien zur Berechnung dünner Schalen mittels der Methode der finiten Elemente bei Verwendung nichtkonformer Verschiebungsansätze, Diss. TH Wien, 1979.
12 PIAN, T. H. H., Derivation of Element Stiffness Matrices by Assumed Stress Distribution, AIAA Journal 2, S. 1333–1336 (1964).
13 KARAMANLIDIS, D., Finite-Element-Modelle der nichtlinearen Tragwerkstatik, Kontinuumsmechanische Grundlagen und Variationsmethoden der Theorie endlicher Verschiebungen, Bericht aus dem 1. Institut für Mechanik der TU Berlin, 1979.

Eingereicht am 2. 5. 1980, in revidierter Fassung am 8. 9. 1980

Anschrift: Prof. Dr. K. KNOTHE, Technische Universität Berlin, Fachbereich 12 — Verkehrswesen, Institut für Luft- und Raumfahrt, Salzufer 17–19, D-1000 Berlin 10

ZAMM 61, 567 – 581 (1981)

B. ZASTRAU / H. ROTHERT

Herleitung einer Direktortheorie für Kontinua mit lokalen Krümmungseigenschaften

Herrn o. Prof. Dr.-Ing. SIGURD FALK zur Vollendung seines 60. Lebensjahres am 6. 5. 1981 gewidmet.

In der vorliegenden Arbeit werden die theoretischen Grundlagen für die Berechnung orientierter Werkstoffe im Rahmen einer verallgemeinerten Cosserat theorie bereitgestellt. Mit der Einführung vektorwertiger Direktoren, die auch eine lokale Änderung der Strukturkrümmung erfassen, werden unter Beachtung der Hauptsätze der Thermodynamik die Feldgleichungen für Massenerhalt, Gleichgewicht und Symmetrie des Direktorspannungstensors hergeleitet. Zur Erfassung der erweiterten kinematischen Möglichkeiten des eingeführten Direktormodells wird ein entsprechend verändertes Grundkonzept der Mechanik vorgestellt. Mit den konstitutiven Beziehungen für ein einfachst mögliches Direktorkontinuum gelingt es ohne Einschränkung der Allgemeingültigkeit, die dargelegte Theorie für die Anwendung mit Finiten Elementen aufzubereiten. Auf der Grundlage von numerischen Ergebnissen erfolgt abschließend eine Diskussion der Bedeutung der eingeführten Direktoreigenschaften.

The investigations that follow aim at the improvement of the theoretical foundation for the numerical treatment of orientated media in the range of generalized Cosserat continua. Introducing vector-valued directors, which are suitable to represent the local change of structural curvature, and applying the two laws of thermodynamics the field equations for the conservation of mass, for the state of equilibrium and for the symmetry of the director stress tensor are being derived. This is achieved by introducing novel aspects of director representation. Introducing then most simple constitutive relations the theory becomes applicable for finite element methods without loss of generality. On the basis of numerical results characteristic features of this type of a generalized Cosserat theory are finally being discussed.

Настоящая рабо тапосвящена теоретическим основам для рассчёта ориентированных материалов в рамках обобщённой теории Коссерата. Вводя направляющие [directors], которые имеют значение векторов и охватывают также локальное изменение кривизны структуры, с помощью основных законов термодинамики выводятся уравнения полей для сохранения массы, равновесия и симметрии тензоров напряжения направляющих. С целью учёта расширенных кинематических возможностей введённой модели направляющей представлены соответственно изменённые основные положения механики. С помощью основополагающих соотношений для наиболее простого континуума направляющих удаётся, не ограничивая общности, подготовить данную теорию для приложения метода конечных элементов. На основе численных результатов обсуждается в заключении значение введённых свойств направляющих.

1. Einleitung

Trotz der offensichtlichen Erfolge, die der Einsatz der EDV und der Methode der Finiten Elemente auf dem Gebiet der mathematisch-numerischen Behandlung von Problemen der Kontinuumsmechanik bewirkt hat, wird in dieser Arbeit der Versuch unternommen, das *theoretische* Konzept der Berechnung orientierter Kontinua zu erweitern. Daß die vorgelegte Theorie schließlich ebenfalls in eine Form überführt wird, die eine Lösung durch Minimierung der inneren Energie liefert, liegt in der Überzeugung der Autoren begründet, daß durch Verwendung finiter Elemente eine Übertragung der allgemeinen Feldgleichungen auf in der Praxis vorzufindende Aufgabenstellungen am einfachsten durchzuführen ist.

Ausgehend von der klassischen Modellvorstellung der Kontinuumsmechanik für Festkörper wird jedem seiner materiellen Punkte zusätzlich die Eigenschaft einer beliebigen Zahl unabhängig voneinander deformierbarer Direktoren zugeordnet. Bei der Einführung von Direktoren sind die beiden Aspekte

— Festlegung der Kinematik ihrer Bewegung und
— ihr Beitrag zur kinetischen Energie des bewegten Körpers

von Bedeutung. Letzteres ist insofern wichtig, als die Herleitung thermodynamisch zulässiger Materialbeschreibungen nur unter Beachtung der Energieerhaltung und der Entropieproduktion möglich ist.

Direktoren sind, ähnlich wie der Drehvektor bei der COSSERATtheorie, geeignet, die Änderung sowohl einer Strukturkrümmung als auch die der Orientierung eines Partikels zu erfassen. Allerdings ermöglicht erst die Berücksichtigung der *Längsverformbarkeit*, einen im Gegensatz zu *polaren* Theorien großräumigen Einfluß der Werkstofforientierung nachzuweisen. Dennoch ist zu beachten, daß zur vollständigen Formulierung der Feldgleichungen für das Kräftegleichgewicht und die innere Energie von der Vorstellung der Invarianz des Beanspruchungszustands bei Überlagerung einer zeitlich konstanten, lokalen Krümmungsgeschwindigkeit auszugehen ist.

Für diese beiden Erweiterungen der klassischen Mechanik werden im folgenden die Gleichgewichtsbeziehungen und ein einfaches Materialgesetz hergeleitet sowie zum Nachweis der Anwendbarkeit am Beispiel der schubbelasteten Scheibe eine FEM-Lösung vorgestellt.

2. Modellvorstellung eines allgemeinen n-Direktorkontinuums

Der Herleitung einer Direktortheorie soll ihr Zusammenhang mit der klassischen Kontinuumsmechanik vorangestellt werden. Üblicherweise werden die Gleichgewichtsbedingungen für einen ausgedehnten Körper aus dem *Kräfte-* und *Momenten*gleichgewicht am differentiellen Element gewonnen. Der Übergang von Spannungen zu Kräften läßt sich

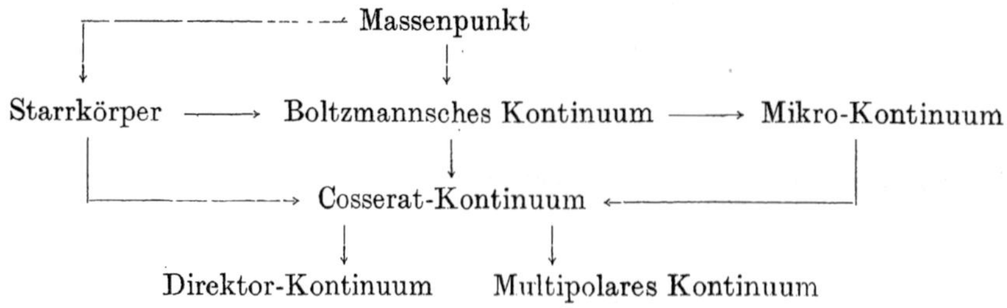

Bild 1. Feldtheorien der Mechanik

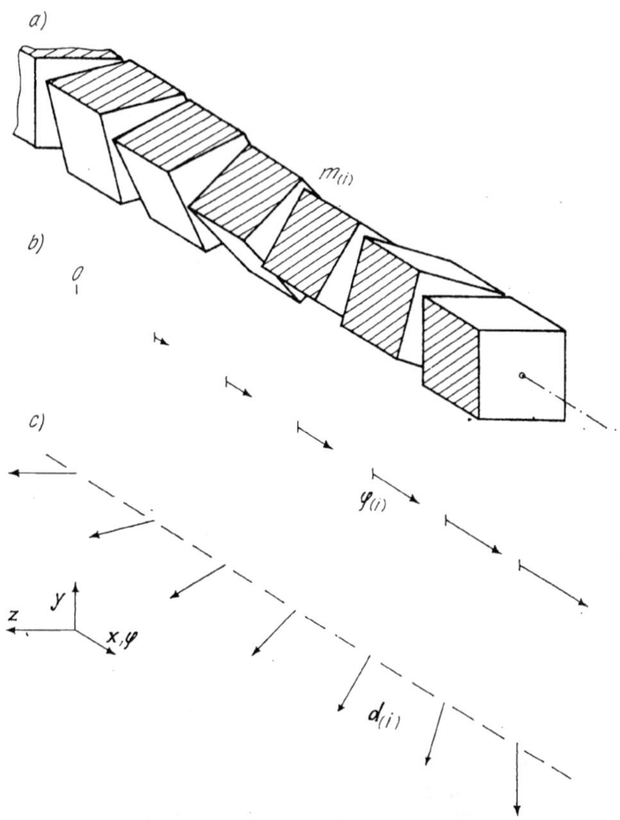

Bild 2. Darstellungsmöglichkeiten eines Verdrehungsfreiheitsgrades

deshalb einfach vollziehen, weil wegen der Kleinheit des Kontinuumsausschnitts die Spannungsverteilung an den Elementrändern als konstant angesehen werden kann. Ein solches Vorgehen entspricht der Auflösung des Kontinuums in eine unendliche Anzahl von Massenpunkten. Zur Formulierung des Momentengleichgewichts wird darüber hinaus von der Körpereigenschaft des infinitesimalen Elements Gebrauch gemacht. Wie in Bild 1 dargestellt, lassen sich daraus unmittelbar zwei Erweiterungen ableiten:

— Versieht man ein Partikel mit den Eigenschaften eines Kontinuums, d. h. gestattet man im Kleinen eine von der Makrodeformation unabhängige Deformation, so entsteht ein Körpermodell, das von ERINGEN [5], MINDLIN [10], STOJANOVIĆ [15] u. a. mit den Begriffen *Mikrokontinuum*, *mikropolare* Theorie bzw. Material mit *Mikrostruktur* eingeführt worden ist.
— Beschränkt man die zusätzlichen lokalen Deformationen auf Starrkörperdrehungen des infinitesimalen Elements, so gelangt man über die Einführung eines unabhängigen Rotationsvektors zur Modellvorstellung der COSSERATtheorie. Die Unterscheidung von Ortskoordinaten und Mikrokoordinaten wie bei den mikropolaren Theorien entfällt.

Ebenfalls unter Verzicht auf die Vorstellung eines Ineinandergeschachteltseins der Kontinua sind von NAGHDI [11] und ROTHERT [14] *Direktortheorien* sowie von GREEN/RIVLIN [7] eine *multipolare* Theorie vorgestellt worden. Während ein multipolarer Körper dadurch gekennzeichnet ist, daß zur Beschreibung des Spannungszustands und der Deformation Tensoren höherer als zweiter Stufe — für ein n-polares Medium $2 + n$-ter Stufe — benötigt werden, erhöht sich bei einer n-Direktortheorie die Zahl der unabhängigen Verschiebungsfreiheitsgrade i. a. mit jedem zusätzlichen Direktor um drei.

Für beide letztgenannten Beschreibungsmöglichkeiten ist festzustellen, daß sie bei geeigneter Umschreibung die zuvor angesprochenen Theorien als Sonderfall enthalten. Für die Direktortheorie möge dieser Sachverhalt anhand der Bilder 2 und 3 kurz aufgezeigt werden. Gleichzeitig soll versucht werden, einen ersten anschaulichen Eindruck von den Aussagemöglichkeiten einer Direktortheorie zu vermitteln.

Nr.	Beschränkungen	Theorie	Charakteristika
1		allgemeine Direktortheorie	Direktorzahl unbegrenzt (n) Direktordeformation beliebig
2	Reduktion der Direktorzahl Starrheit des einzelnen Direktors	allgemeine COSSERATtheorie	Direktorzahl $n = 3$ Direktoren drehbar Partikelverschiebung beliebig
3	Starrheit der Direktoren untereinander	COSSERATtheorie: trièdre caché	Einführung eines Rotationsvektors Partikelverschiebung beliebig und kompatibel Momentenspannungen (couple stresses) Kraftspannungen
4	Beschränkung der Rotation	COSSERATtheorie: constrained rotation	Direktordrehung entspricht der Rotation des Verschiebungsfeldes Partikelverschiebung beliebig Momentenspannungen Kraftspannungen
5	Material ist weich in Bezug auf COSSERATeffekte	BOLTZMANNkontinuum	Partikelverschiebung beliebig Kraftspannungen
6	Verzerrungen verschwinden	Starrkörper	Kräfte und Momente

Bild 3. Hierarchie einer Theorieerweiterung

Als naheliegendstes Beispiel bietet sich die allgemeine *Stabtheorie* an. Während die Phänomene der Biegung und Dehnung allein durch die Kenntnis der Lage der Punkte der Stabachse vor und nach der Deformation hinreichend beschreibbar sind, muß zur Erfassung der Torsion ein zusätzlicher Freiheitsgrad eingeführt werden, der die Änderung der Orientierung der Struktur wiedergibt. Neben der Richtungsänderung (Verdrehung der Stabachse) kann durch die Deformation des Direktors bei geeigneten konstitutiven Gesetzen gleichzeitig eine Aussage über z. B. die Dickenänderung oder die Querschnittsverwölbung des Stabes getroffen werden. Auf eine weitere Ausformulierung einer Direktor*stab*theorie soll hier angesichts des zu behandelnden *Kontinuums*problems verzichtet werden.

Einen abschließenden Überblick über die Einbindung der Direktortheorie in die klassische Konzeption der Mechanik gewinnt man aus Bild 3.

Die hier verwendete Direktortheorie geht zunächst in Anlehnung an NAGHDI [11] von folgenden Voraussetzungen aus: Jedem materiellen Punkt eines Körpers können n voneinander unabhängige Direktoren ${}^i\boldsymbol{d}$ mit $1 \leqq i \leqq n$ zugeordnet werden, wobei auch für $i \leqq 3$ die Direktoren im Ausgangszustand ${}^i\boldsymbol{D}$ weder normiert noch linear unabhängig voneinander zu sein brauchen. Zur Lösung des Verformungsproblems müssen außer dem Verschiebungsfeld $\boldsymbol{u}(t)$, das den Ortsvektor $\boldsymbol{r}(t)$ zur Zeit t mit der Ausgangskonfiguration $\boldsymbol{R}(\theta^j) = \boldsymbol{r}(\theta^j, t = t_0)$ verbindet, die entsprechenden Direktorverschiebungsfelder ${}^i\boldsymbol{u}(\theta^j, t)$ bekannt sein. Dabei wird angenommen, daß die Direktorverschiebung von einer Starrkörpertranslation nicht beeinflußt wird (zur bildlichen Darstellung dieser Modellvorstellung vgl. Bild 4).

Die Unabhängigkeit der Direktoren untereinander sowie von der Translationsgeschwindigkeit des Körpers kommt auch in der Formulierung des Energieerhaltungssatzes zum Ausdruck:

$$\frac{D}{Dt}\int_v \left\{\frac{1}{2}\varrho\left(\boldsymbol{v}\cdot\boldsymbol{v} + \sum_{i=1}^{n} {}^i\alpha\, {}^i\boldsymbol{v}\cdot {}^i\boldsymbol{v}\right) + \varrho U\right\} \mathrm{d}V =$$

$$= \int_v \varrho\left\{\boldsymbol{v}\cdot\boldsymbol{f} + \sum_{i=1}^{n} {}^i\alpha\, {}^i\boldsymbol{f}\cdot {}^i\boldsymbol{v} + r\right\}\mathrm{d}V + \int_o \left\{\boldsymbol{v}\cdot\boldsymbol{s} + \sum_{i=1}^{n} {}^i\boldsymbol{v}\cdot {}^i\boldsymbol{s} - h\right\}\mathrm{d}O\,. \tag{2.1}$$

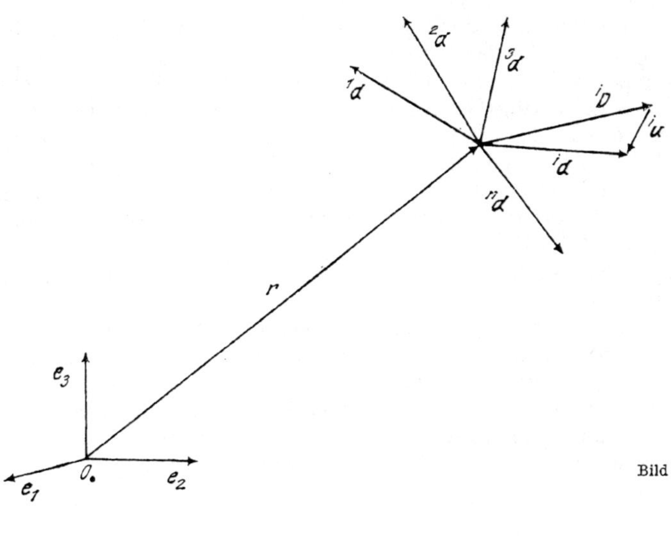

Bild 4. Graphische Darstellung eines Direktorkontinuums

Es bedeuten D/Dt die materielle Zeitableitung, ϱ die Dichte, ${}^i\alpha$ ein zeitlich konstanter Beiwert zur Dichte für den Beitrag der Direktorgeschwindigkeit ${}^i\boldsymbol{v}$ des i-ten Direktors zur kinetischen Energie, $\boldsymbol{v}$ die Geschwindigkeit des Massenpunktes, $\boldsymbol{f}$ und ${}_i\boldsymbol{f}$ massebezogene Volumenkräfte bzw. Direktorkräfte, $\boldsymbol{s}$ und ${}_i\boldsymbol{s}$ oberflächenbezogene Schnittkräfte bzw. Direktorschnittkräfte, U die spezifische innere Energie, r eine skalare Wärmequelle sowie h den Wärmefluß pro Fläche. Die Bezeichnungen $\mathrm{d}V$, $\mathrm{d}O$ sowie V und O stehen stellvertretend für ein differentielles Volumen- oder Oberflächenelement bzw. gelten als Hinweis für die Integration über das Volumen oder die Oberfläche des Körpers.

Obwohl im Energiesatz die mit den Direktoreigenschaften verknüpften Ausdrücke additiv erscheinen, erhält man über den Einfluß der Starrkörperrotation auf den Bewegungszustand der Direktoren eine Kopplung der Schnittkräfte über die zweite CAUCHYsche Bewegungsgleichung. Einzelheiten hierzu sind der in Kap. 4 ausgeführten Herleitung der Feldgleichungen zu entnehmen.

Im übrigen wird im folgenden von der Summationskonvention Gebrauch gemacht, wonach bei doppelt auftretenden lateinischen (griechischen) Indizes bis 3(2) summiert wird. Komponenten gelten, soweit nichts anderes vereinbart wird, bezüglich der jeweils aktuellen Basis $\boldsymbol{g}_i$.

3. Überlegungen zur Einführung eines erweiterten kinematischen Modells

Die Herleitung der Gleichgewichtsbeziehungen eines Direktorkontinuums geschieht zweckmäßigerweise unter Ausnutzung der beiden Hauptsätze der Thermodynamik und der Forderung, daß die innere Energie eines Körpers unabhängig von einer dem aktuellen Deformationszustand überlagerten *Starrkörperbewegung* sein muß. Dabei führt die Mitnahme einer unbeschleunigten Translationsgeschwindigkeit zum Kräftegleichgewicht und die Berücksichtigung einer Rotation zum Momentengleichgewicht. Bei einer klassischen Beschreibung des Kontinuums erhält man daraus die Symmetrie des Spannungstensors. Beim Direktorkontinuum gewinnt man hingegen nur eine Verknüpfung zwischen dem antimetrischen Anteil des Spannungstensors und einem Ausdruck aus der Direktorbelastung und dem entsprechenden Direktorspannungstensor. Es ist naheliegend, die möglichen Bewegungszustände eines Direktorkontinuums in gleicher Weise zu erweitern, wie dies durch die Einführung der Rotation für einen Körper geschehen ist, dessen Bausteine nur bei Verschiebungen in der Lage sind, Arbeit zu verrichten, nicht aber bei Drehungen. Eine derartige Ergänzung muß folgende Forderungen erfüllen:

— Entlang ihrer Wirkungsrichtung dürfen Kräfte und Momente keine Arbeit leisten; umgekehrt müssen die zugehörigen Kraftgrößen bei Verschiebungen und Verdrehungen ohne Arbeitsanteil bleiben.
— Die hinzugewonnenen Weg- bzw. Kraftgrößen müssen auf elementare Weise durch die bekannten beschreibbar sein.

Beide Bedingungen werden bei Verwendung einer *lokalen Strukturkrümmung* erfüllt, wie es anhand eines Dimensionsvergleichs unmittelbar festzustellen ist. Ebenso wie bei kleinen Rotationen über das Kreuzprodukt des Ortsvektors $\boldsymbol{r}$ mit dem Drehvektor $\boldsymbol{\varphi}$ die zugehörige Verschiebung $\boldsymbol{u}$ eines Punktes bestimmt werden kann, läßt sich in gleicher Weise aus dem Vektor der Verkrümmung $\boldsymbol{\varkappa}$ eine entsprechende Verdrehung $\boldsymbol{\varphi}_\varkappa$ bestimmen. Es gelte

$$\boldsymbol{u} = \boldsymbol{\varphi} \times (\boldsymbol{r} - \boldsymbol{r}_0), \qquad \boldsymbol{\varphi}_\varkappa = \boldsymbol{\varkappa} \times (\boldsymbol{r} - \boldsymbol{r}_0), \tag{3.1}$$

wobei mit $\boldsymbol{r}_0$ der Ortsvektor des Punktes bezeichnet sei, der als Fixpunkt der Rotation bzw. Zentrum der Krümmung anzusehen ist. Die Erweiterung der verwendeten Kraftgrößen kann in analoger Weise durchgeführt werden:

$$\boldsymbol{H} = (\boldsymbol{r} - \boldsymbol{r}_0) \times \boldsymbol{M}, \qquad \boldsymbol{M} = (\boldsymbol{r} - \boldsymbol{r}_0) \times \boldsymbol{F}. \tag{3.2}$$

$\boldsymbol{F}$ und $\boldsymbol{M}$ sind die Bezeichnungen für Kraft und Moment, während für $\boldsymbol{H}$ der Begriff *Hypermoment* in Anlehnung an das gleiche Bildungsgesetz für $\boldsymbol{M}$ und $\boldsymbol{H}$ neu eingeführt wird. Der Zusammenhang von $\boldsymbol{H}$ und $\boldsymbol{\varkappa}$ mit den klassisch verwendeten Kraft- und Weggrößen wird auch aus der Beziehung für die Verschiebungsarbeit W deutlich:

$$W = \boldsymbol{\varkappa} \cdot \boldsymbol{H} = \boldsymbol{\varkappa} \cdot \{(\boldsymbol{r} - \boldsymbol{r}_0) \times [(\boldsymbol{r} - \boldsymbol{r}_0) \times \boldsymbol{F}]\} = \{[\boldsymbol{\varkappa} \times (\boldsymbol{r} - \boldsymbol{r}_0)] \times (\boldsymbol{r} - \boldsymbol{r}_0)\} \cdot \boldsymbol{F}. \tag{3.3}$$

Obwohl die Umformulierung von (3.3) aus der Sicht der Vektoralgebra fehlerfrei ist, unterliegt die stillschweigende Verwendung von (3.1) einer Einschränkung hinsichtlich der Größe von $\boldsymbol{\varphi}$ bzw. von $\boldsymbol{\varkappa}$. Im Rahmen einer linearen Näherung hat (3.1) nur Gültigkeit, wenn $\boldsymbol{\varphi}$ und $\boldsymbol{\varkappa}$ klein bzw. infinitesimal sind. Faßt man (3.1) zu einer einzigen Beziehung zusammen, so muß darüber hinaus *Lokalität*, d. h. Beschränktheit der Differenz zwischen $\boldsymbol{r}$ und $\boldsymbol{r}_0$ verlangt werden. Demgegenüber sind entsprechende Forderungen zur Bestimmung von $\boldsymbol{H}$ nicht erforderlich. Da auch für die Herleitung einer Direktortheorie alle Feldgleichungen streng nur im verformten System erfüllt sind und eine Änderung des Zustands stets durch die aktuellen Geschwindigkeiten beschrieben wird, ist die Forderung nach Lokalität von $\boldsymbol{\varkappa}$ und Infinitesimalität von $\boldsymbol{\varphi}$ nicht als Einschränkung der allgemeinen Theorie zu werten, sondern erfordert für große Deformationen lediglich eine inkrementelle Vorgehensweise.

Die Bedeutung des eingeführten Hypermoments läßt sich in Erweiterung von ZASTRAU [16] auch durch Anwendung auf klassische Kontinuumsprobleme erkennen. Zur Ermittlung der Gleichgewichtsbeziehungen werden im Rahmen der klassischen Mechanik die Spannungen zu Kräften zusammengefaßt und verlangt, daß ihre Resultierende verschwindet. Entsprechend der Bildungsvorschrift für das Moment als Kreuzprodukt aus Kraft und Ortsvektor muß eine mögliche Momentenbelastung in Richtung der Flächennormalen eingetragen werden. Verlangt man nunmehr auch Gleichgewicht hinsichtlich der neu eingeführten Kraftgröße, so kann nach Verzicht auf eine äußere Momentenbelastung auf die Gleichheit der Schubspannungen und bei Übertragung auf das Kontinuum auf die Symmetrie des Spannungstensors geschlossen werden. Daß zum Moment eine Verdrehung als zugeordnete Bewegung gehört, ist bereits aus der Stabtheorie bekannt.

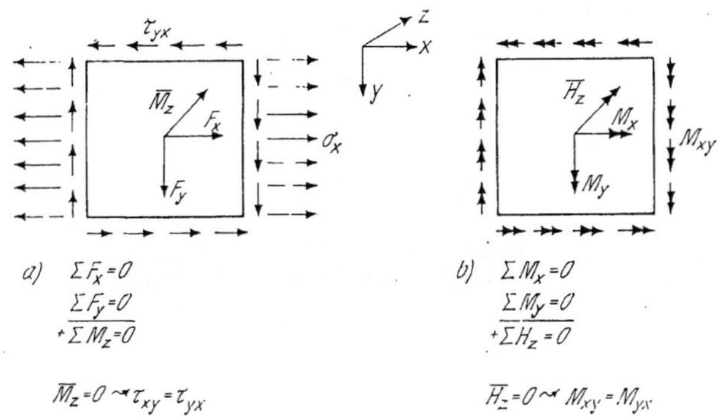

Bild 5. Scheibe unter Kräfte- und Momentenbeanspruchung

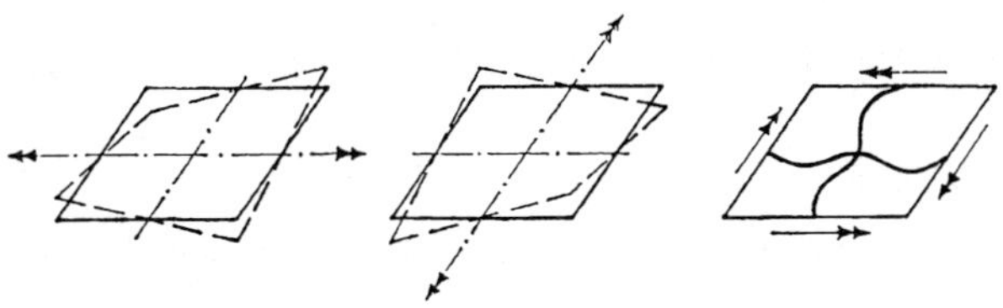

Bild 6. Gedachte Flächenkrümmung infolge der Schnittmomente

In Bild 5b werden qualitativ dieselben Schritte für ein durch Momente belastetes Scheiben-/Plattenelement durchgeführt. Auch hier kann nach Einführung des Hypermoments durch Verzicht auf eine entsprechende äußere Belastung auf die Symmetrie des zugehörigen Momententensors geschlossen werden.

Die Frage nach der zugehörigen Bewegung wird durch Bild 6 anschaulich beantwortet. Gleichzeitig erkennt man daran die lineare Unabhängigkeit von Kraft- und Weggrößen unterschiedlicher Stufe. Eins zeigt dieses Bild jedoch auch, daß nämlich für eine vollständige Beurteilung der gegenseitigen Beeinflussung von z. B. Flächenkrümmung und Spannungsverteilung die Mitnahme des kompletten Spannungstensors und nicht nur eines ebenen Ausschnitts wichtig ist. Durch die Forderung nach Gleichheit von M_{xy} und M_{yx} kann erreicht werden, daß σ_z jeweils an einer Oberfläche des dargestellten Plattenelements keine Arbeit leistet.

Dieser kurze Versuch einer anschaulichen Darstellung möge deutlich gemacht haben, daß *Hypermoment* und *Raumkrümmung* im Sinne der zu Beginn von Kap. 3 aufgestellten Forderungen wirklich als *unabhängige* Größen eingeführt wurden und durchaus geeignet sind, *eine Mechanik erweiterter Freiheitsgrade* zu beschreiben.

Für die weiteren Schritte der Herleitung einer Direktortheorie sind bei Anwendung des hier vorgestellten Konzepts zwei zusätzliche Vereinbarungen erforderlich:

— Die innere Energie eines Körpers ändert sich nicht bei Überlagerung einer aus der eingeführten Krümmung $\boldsymbol{\varkappa}$ abgeleiteten räumlich und zeitlich konstanten Krümmungsgeschwindigkeit $\boldsymbol{K}$.

— Die der Raumkrümmung zuzuordnenden Kraftgrößen können als nicht vorhanden vorausgesetzt werden.

4. Grundgleichungen einer Direktortheorie bei Gültigkeit erweiterter Invarianzforderungen

Die Herleitung einer neuartigen Kontinuumstheorie muß sich auf zweifelsfrei formulierbare Grundbeziehungen abstützen. Aus dem Bereich der Mechanik möge dies an erster Stelle der entsprechend (2.1) erweiterte Energiehaltungssatz sein. Weiterhin muß gefordert werden, daß auch für ein Direktorkontinuum die Entropieproduktionsungleichung erfüllt sei. Ohne auf die zur Zeit geführte Diskussion eingehen zu wollen, welche Formulierung als die zweckmäßigste anzusehen ist, wird die Beziehung von Clausius/Duhem in folgender Form verwendet:

$$\frac{D}{Dt}\int\limits_{V} \varrho S \,\mathrm{d}V \geqq \int\limits_{V} \varrho \frac{r}{T} \,\mathrm{d}V - \int\limits_{O} \frac{h}{T} \,\mathrm{d}O \,. \tag{4.1}$$

Es sind S die Entropie und T die absolute Temperatur. Damit bei Annahme eines stetigen Verlaufs aller Lösungen die Anwendung des Gaussschen Satzes

$$\int\limits_{O} \mathbf{T} \cdot \boldsymbol{n} \,\mathrm{d}O = \int\limits_{V} \nabla \cdot \mathbf{T}^T \,\mathrm{d}V \tag{4.2}$$

zur Überführung der integralen Beziehungen in lokale Feldgleichungen möglich wird, sei vorausgesetzt, daß nach Einführung einer Flächennormalen $\boldsymbol{n}$ sowohl der Spannungstensor $\mathbf{T}$ als auch der Direktorspannungstensor ${}_i\mathbf{T}$ über

$$\mathbf{T} \cdot \boldsymbol{n} = \boldsymbol{s} \,, \qquad {}_i\mathbf{T} \cdot \boldsymbol{n} = {}_i\boldsymbol{s} \tag{4.3 a, b}$$

eingeführt werden können. In gleicher Weise werde der Zusammenhang zwischen dem Wärmeflußvektor $\boldsymbol{h}$ und dem gemäß (2.1) eingeführten Wärmefluß h hergestellt.

Weitere Voraussetzung für eine allgemeine Direktortheorie sei die Forderung, daß die innere Energie ϱU eines Körpers auch bei Vorhandensein eines Direktordeformationsfeldes unabhängig von der im Deformationszustand enthaltenen Starrkörperbewegung bleibt. Neben der bereits eingeführten Unabhängigkeit der kinetischen Energie und der Leistung der Direktorbewegung vom sonstigen Geschwindigkeitsfeld des Körpers ist das Transformationsverhalten der ${}^i\boldsymbol{v}$ bei Überlagerung einer Starrkörperbewegung vorzugeben. Es gelte:

$$\bar{\boldsymbol{v}} = \boldsymbol{v} + \boldsymbol{b}\,, \qquad {}^i\bar{\boldsymbol{v}} = {}^i\boldsymbol{v}\,, \tag{4.4 a, b}$$

$$\hat{\boldsymbol{v}} = \boldsymbol{v} + \boldsymbol{\Omega} \times (\boldsymbol{r} - \boldsymbol{r}_0)\,, \qquad {}^i\hat{\boldsymbol{v}} = {}^i\boldsymbol{v} + \boldsymbol{\Omega} \times {}^i\boldsymbol{d}\,, \tag{4.5 a, b}$$

$$\overset{*}{\boldsymbol{v}} = \boldsymbol{v} + \{\boldsymbol{K} \times (\boldsymbol{r} - \boldsymbol{r}_0)\} \times (\boldsymbol{r} - \boldsymbol{r}_0)\,, \qquad {}^i\overset{*}{\boldsymbol{v}} = {}^i\boldsymbol{v} + \{\boldsymbol{K} \times (\boldsymbol{r} - \boldsymbol{r}_0)\} \times {}^i\boldsymbol{d}\,. \tag{4.6 a, b}$$

In diesen Beziehungen sind $\boldsymbol{b}$, $\boldsymbol{\Omega}$ und $\boldsymbol{K}$ jeweils zeitlich und örtlich konstante Geschwindigkeiten der Translation, Rotation und Raumkrümmung. Führt man unter Beachtung der Lokalität von $\boldsymbol{K}$ für die Differenz der Ortsvektoren

$$\delta\boldsymbol{r} = \boldsymbol{r} - \boldsymbol{r}_0\,, \qquad |\delta\boldsymbol{r}| \ll 1 \tag{4.7 a, b}$$

ein, so ergibt sich aus (4.6) nach Verzicht auf die in $\delta\boldsymbol{r}$ quadratischen Terme

$$\overset{*}{\boldsymbol{v}} \approx \boldsymbol{v}\,, \qquad {}^i\overset{*}{\boldsymbol{v}} = {}^i\boldsymbol{v} + (\boldsymbol{K} \times \delta\boldsymbol{r}) \times {}^i\boldsymbol{d}\,. \tag{4.8 a, b}$$

Damit wird einerseits die Direktorbewegung als unabhängig von einer Translation und die klassisch verwendete Geschwindigkeit andererseits als näherungsweise unabhängig von der Raumkrümmung aufgefaßt. Die Beziehungen für Massenerhalt und Gleichgewicht folgen unmittelbar aus der Forderung:

$$\varrho U(\boldsymbol{v}, {}^i\boldsymbol{v}) = \varrho U(\bar{\boldsymbol{v}}, {}^i\bar{\boldsymbol{v}}) = \varrho U(\hat{\boldsymbol{v}}, {}^i\hat{\boldsymbol{v}}) = \varrho U(\overset{*}{\boldsymbol{v}}, {}^i\overset{*}{\boldsymbol{v}})\,. \tag{4.9}$$

Verwendet man nacheinander die mit $(\bar{\ldots})$, $(\hat{\ldots})$ und $(\overset{*}{\ldots})$ bezeichneten Bewegungszustände zur Formulierung des Energiesatzes

$$\frac{D}{Dt}\int_V \varrho\left\{\frac{1}{2}\bar{\boldsymbol{v}}\cdot\bar{\boldsymbol{v}} + \sum_{i=1}^{n}\frac{{}^i\alpha}{2}\,{}^i\bar{\boldsymbol{v}}\cdot{}^i\bar{\boldsymbol{v}} + U\right\}\mathrm{d}V$$
$$= \int_V \varrho\left\{\bar{\boldsymbol{v}}\cdot\boldsymbol{f} + \sum_{i=1}^{n}{}^i\alpha\,{}_i\boldsymbol{f}\cdot{}^i\boldsymbol{v} + r\right\}\mathrm{d}V + \int_O\left\{\bar{\boldsymbol{v}}\cdot\boldsymbol{s} + \sum_{i=1}^{n}{}_i\boldsymbol{s}\cdot{}^i\bar{\boldsymbol{v}} - h\right\}\mathrm{d}O \tag{4.10}$$

und bildet die Differenz zur Ausgangsform (2.1), so folgt:

$$\frac{D}{Dt}\int_V \varrho\left(\boldsymbol{v}\cdot\boldsymbol{b} + \frac{1}{2}\boldsymbol{b}\cdot\boldsymbol{b}\right)\mathrm{d}V = \int_V \varrho\boldsymbol{b}\cdot\boldsymbol{f}\,\mathrm{d}V + \int_O \boldsymbol{b}\cdot\boldsymbol{s}\,\mathrm{d}O\,, \tag{4.11}$$

$$\frac{D}{Dt}\int_V \varrho\left\{\frac{1}{2}\left[2\boldsymbol{v}\cdot(\boldsymbol{\Omega}\times\boldsymbol{r}) + (\boldsymbol{\Omega}\times\boldsymbol{r})\cdot(\boldsymbol{\Omega}\times\boldsymbol{r}) + \sum_{i=1}^{n}{}^i\alpha\,\langle 2\,{}^i\boldsymbol{v}\cdot(\boldsymbol{\Omega}\times{}^i\boldsymbol{d}) + (\boldsymbol{\Omega}\times{}^i\boldsymbol{d})\cdot(\boldsymbol{\Omega}\times{}^i\boldsymbol{d})\rangle\right]\right\}\mathrm{d}V$$
$$= \int_V \varrho\left\{(\boldsymbol{\Omega}\times\boldsymbol{r})\cdot\boldsymbol{f} + \sum_{i=1}^{n}{}^i\alpha(\boldsymbol{\Omega}\times{}^i\boldsymbol{d})\cdot{}_i\boldsymbol{f}\right\}\mathrm{d}V + \int_O\left\{(\boldsymbol{\Omega}\times\boldsymbol{r})\cdot\boldsymbol{s} + \sum_{i=1}^{n}(\boldsymbol{\Omega}\times{}^i\boldsymbol{d})\cdot{}_i\boldsymbol{s}\right\}\mathrm{d}O\,, \tag{4.12}$$

$$\int_V \varrho\sum_{i=1}^{n}{}^i\alpha\left\{[{}^i\dot{\boldsymbol{v}} + (\boldsymbol{K}\times\delta\dot{\boldsymbol{r}})\times{}^i\boldsymbol{d} + (\boldsymbol{K}\times\delta\boldsymbol{r})\times{}^i\boldsymbol{v}]\cdot[{}^i\boldsymbol{v} + (\boldsymbol{K}\times\delta\boldsymbol{r})\times{}^i\boldsymbol{d}] - {}^i\dot{\boldsymbol{v}}\cdot{}^i\boldsymbol{v}\right\}\mathrm{d}V$$
$$= \int_V \varrho\left\{\sum_{i=1}^{n}{}^i\alpha[(\boldsymbol{K}\times\delta\boldsymbol{r})\times{}^i\boldsymbol{d}]\cdot{}_i\boldsymbol{f}\right\}\mathrm{d}V + \int_O\sum_{i=1}^{n}[(\boldsymbol{K}\times\delta\boldsymbol{r})\times{}^i\boldsymbol{d}]\cdot{}_i\boldsymbol{s}\,\mathrm{d}O\,. \tag{4.13}$$

Dabei wurde zur Vermeidung unnötiger Schreibarbeit in (4.13) bereits von der Beziehung für den Massenerhalt Gebrauch gemacht. Sie folgt aus (4.11) dadurch, daß wegen der Beliebigkeit von $\boldsymbol{b}$ die reduzierte Beziehung

$$\frac{D}{Dt}\int_V \varrho\left(\boldsymbol{v} + \frac{1}{2}\boldsymbol{b}\right)\mathrm{d}V = \int_V \varrho\boldsymbol{f}\,\mathrm{d}V + \int_O \boldsymbol{s}\,\mathrm{d}O \tag{4.14}$$

in zwei Teilaussagen aufgespalten werden kann:

$$\frac{D}{Dt}\int_V \varrho\,\mathrm{d}V = 0\,, \qquad \frac{D}{Dt}\int_V \varrho\boldsymbol{v}\,\mathrm{d}V = \int_V \varrho\boldsymbol{f}\,\mathrm{d}V + \int_O \boldsymbol{s}\,\mathrm{d}O\,. \tag{4.15), (4.16}$$

Nach Anwendung des GAUSSschen Integralsatzes ergeben sich in lokaler Form

$$\dot{\varrho} + \varrho\nabla\cdot\boldsymbol{v} = 0\,, \qquad \varrho\dot{\boldsymbol{v}} = \varrho\boldsymbol{f} + \nabla\cdot\mathbf{T}^T\,. \tag{4.17), (4.18}$$

Hierbei entsprechen der übergesetzte Punkt der materiellen Zeitableitung und das ∇ Symbol dem Nablaoperator. Im übrigen zeigt sich, daß Massenerhalt und Kräftegleichgewicht eines Direktorkontinuums keinerlei Unterschied zum klassischen Kontinuum aufweisen.

Die Auswertung der zweiten Restgleichung (4.12) bedarf entweder eines Kunstgriffs oder einer Einschränkung dahingehend, daß nur infinitesimale Rotationsgeschwindigkeiten zulässig sind. Beides hat den Zweck, Terme zu eliminieren, die durch Aufgabe der Inertialeigenschaft des Bezugssystems die Mitnahme von Führungsbeschleunigungen und entsprechenden Trägheitskräften erforderlich machen würden.

Anders als bei BUGGISCH [4] und ERINGEN [5], die den Übergang zu rotierenden Systemen auch dazu benötigen, Gleichungen für den Rotationsträgheitserhalt zu bestimmen, kann mit (4.15) und der getroffenen Vereinbarung

$$ {}^i\dot{\alpha} = 0 \tag{4.19} $$

Gleichung (4.12) unmittelbar zur Bestimmung der zweiten CAUCHYschen Bewegungsgleichung herangezogen werden. Nach einigen Rechenschritten ergibt sich

$$ \begin{aligned} &\int\limits_v \varrho \Big\{ \boldsymbol{r} \times \dot{\boldsymbol{v}} + \tfrac{1}{2} \langle 2\dot{\boldsymbol{\Omega}}(\boldsymbol{r} \cdot \boldsymbol{v}) - \boldsymbol{v}(\boldsymbol{r} \cdot \boldsymbol{\Omega}) - \boldsymbol{r}(\boldsymbol{v} \cdot \boldsymbol{\Omega}) \rangle + \\ &\quad + \sum_{i=1}^{n} {}^i\alpha [{}^i\boldsymbol{d} \times {}^i\dot{\boldsymbol{v}} + \tfrac{1}{2} \langle 2\boldsymbol{\Omega}({}^i\boldsymbol{d} \cdot {}^i\boldsymbol{v}) - {}^i\boldsymbol{v}({}^i\boldsymbol{d} \cdot \boldsymbol{\Omega}) - {}^i\boldsymbol{d}({}^i\boldsymbol{v} \cdot \boldsymbol{\Omega}) \rangle] \Big\} \, \mathrm{d}V \\ &= \int\limits_v \varrho \Big\{ \boldsymbol{r} \times \boldsymbol{f} + \sum_{i=1}^{n} {}^i\alpha \, {}^i\boldsymbol{d} \times {}_i\boldsymbol{f} \Big\} \, \mathrm{d}V + \int\limits_v \Big\{ \boldsymbol{r} \times \boldsymbol{s} + \sum_{i=1}^{n} {}^i\boldsymbol{d} \times {}_i\boldsymbol{s} \Big\} \, \mathrm{d}O \, . \end{aligned} \tag{4.20} $$

Hätte man zuvor die Erfüllung des Gleichgewichts auch im rotierenden Zustand vorausgesetzt, könnte auf den Gebrauch der Trägheitsmomente

$$ \boldsymbol{m}_z = \boldsymbol{\Omega}(\boldsymbol{r} \cdot \boldsymbol{v}) - \tfrac{1}{2} \boldsymbol{v}(\boldsymbol{r} \cdot \boldsymbol{\Omega}) - \tfrac{1}{2} \boldsymbol{r}(\boldsymbol{v} \cdot \boldsymbol{\Omega}) \, , \tag{4.21a} $$

$$ {}_i\boldsymbol{m}_z = \boldsymbol{\Omega}({}^i\boldsymbol{d} \cdot {}^i\boldsymbol{v}) - \tfrac{1}{2} {}^i\boldsymbol{v}({}^i\boldsymbol{d} \cdot \boldsymbol{\Omega}) - \tfrac{1}{2} {}^i\boldsymbol{d}({}^i\boldsymbol{v} \cdot \boldsymbol{\Omega}) \tag{4.21b} $$

verzichtet werden. Die Anwendung des GAUSSschen Integralsatzes für

$$ \int\limits_0 \boldsymbol{r} \times \mathbf{T} \cdot \boldsymbol{n} \, \mathrm{d}O = \int\limits_v (\boldsymbol{r} \times T^{ik} \boldsymbol{g}_i)_{|k} \, \mathrm{d}V \, , \tag{4.22a} $$

$$ \int\limits_0 {}^i\boldsymbol{d} \times {}_i\mathbf{T} \cdot \boldsymbol{n} \, \mathrm{d}O = \int\limits_v ({}^i\boldsymbol{d} \times {}_iT^{jk} \boldsymbol{g}_j)_{|k} \, \mathrm{d}V \tag{4.22b} $$

führt zur lokalen Form des Momentengleichgewichts. Bei Vernachlässigung der mit $\boldsymbol{\Omega}$ behafteten infinitesimalen Anteile ergibt sich nach teilweiser Ausführung der kovarianten Differentiation $(\ldots)|$ über

$$ \boldsymbol{r}_{,j} = \boldsymbol{g}_j = r^k|_j \, \boldsymbol{g}_k \, , \tag{4.23} $$

$$ \begin{aligned} &\varrho\{\boldsymbol{r} \times \dot{\boldsymbol{v}} - \boldsymbol{r} \times \boldsymbol{f}\} - \boldsymbol{r} \times (T^{jk}|_k \, \boldsymbol{g}_j) + \varrho \sum_{i=1}^{n} {}^i\alpha \, {}^i\boldsymbol{d} \times {}^i\dot{\boldsymbol{v}} \\ &= \varrho \sum_{i=1}^{n} {}^i\alpha \, {}^i\boldsymbol{d} \times {}_i\boldsymbol{f} + r^l|_k \, \boldsymbol{g}_l \times T^{jk} \boldsymbol{g}_j + \sum_{i=1}^{n} ({}^i\boldsymbol{d} \times {}_iT^{jk} \boldsymbol{g}_j)_{|k} \, . \end{aligned} \tag{4.24} $$

Löst man alle mit $\boldsymbol{r} \times$ behafteten Glieder unter Beachtung von (4.18) heraus, so verbleibt eine Beziehung für das Momentengleichgewicht im Direktorkontinuum:

$$ \varrho \sum_{i=1}^{n} {}^i\alpha \, {}^i\boldsymbol{d} \times {}^i\dot{\boldsymbol{v}} = \varrho \sum_{i=1}^{n} {}^i\alpha \, {}^i\boldsymbol{d} \times {}_i\boldsymbol{f} + \sum_{i=1}^{n} ({}^i\boldsymbol{d} \times {}_iT^{jk} \boldsymbol{g}_j)_{|k} + \boldsymbol{g}_k \times T^{jk} \boldsymbol{g}_j \, . \tag{4.25} $$

Die Einführung einer Direktorrestbelastung

$$ {}_i\hat{\boldsymbol{f}} = \varrho \, {}^i\alpha \, {}_i\boldsymbol{f} + {}_iT^{jk}|_k \, \boldsymbol{g}_j - \varrho \, {}^i\alpha \, {}^i\dot{\boldsymbol{v}} \, , \tag{4.26} $$

liefert eine u. a. bei ROTHERT [14] für ein Schalenkontinuum angegebene Symmetriebedingung für die Summe aus dem Kontinuumsspannungstensor $\mathbf{T}$ und dem Direktorspannungstensor ${}_i\mathbf{T}$ verknüpft über den aktuellen Wert des einzelnen Direktors

$$ \boldsymbol{g}_l \times \Big\{ \sum_{i=1}^{n} {}^id^l \, {}_i\hat{\boldsymbol{f}} + \sum_{i=1}^{n} {}^id^l|_k \; {}_iT^{jk} \boldsymbol{g}_j + T^{jl} \boldsymbol{g}_j \Big\} = \boldsymbol{0} \, , \tag{4.27} $$

wobei mit ${}^id^l$, ${}_iT^{jk}$ und T^{jl} die Komponenten der zugehörigen Vektoren bzw. Tensoren verwendet wurden. Diese Beziehung wird später für den Sonderfall des 1-Direktorkontinuums zur Verknüpfung der Direktorrestbelastung mit dem unsymmetrischen Anteil des Spannungstensors $\mathbf{T}$ weiter ausgewertet werden.

Durch Umformung der letzten Variante des Energiesatzes gemäß (4.13) gelangt man zu einer letzten Gleichgewichtsbedingung. Dabei führt die Näherung entsprechend (4.8) wegen (4.7b) nach nochmaliger Vernachlässigung der in $\delta\boldsymbol{r}$ bzw. $\dot{\overline{\delta\boldsymbol{r}}}$ quadratischen Glieder zu

$$ \begin{aligned} &\int\limits_v \varrho \sum_{i=1}^{n} {}^i\alpha \{ (\boldsymbol{K} \times \delta\boldsymbol{r}) \cdot ({}^i\boldsymbol{d} \times {}^i\dot{\boldsymbol{v}}) + (\boldsymbol{K} \times \dot{\overline{\delta\boldsymbol{r}}}) \cdot ({}^i\boldsymbol{v} \times {}^i\boldsymbol{d}) \} \, \mathrm{d}V \\ &= \int\limits_v \varrho \sum_{i=1}^{n} {}^i\alpha (\boldsymbol{K} \times \delta\boldsymbol{r}) \cdot ({}^i\boldsymbol{d} \times {}_i\boldsymbol{f}) \, \mathrm{d}V + \int\limits_o \sum_{i=1}^{n} (\boldsymbol{K} \times \delta\boldsymbol{r}) \cdot ({}^i\boldsymbol{d} \times {}_i\boldsymbol{s}) \, \mathrm{d}O \, . \end{aligned} \tag{4.28} $$

Mit der Einführung eines beliebigen, aber festen Wertes für $\delta\boldsymbol{r}$ kann im Sinne der Lokalität der Raumkrümmung

$$ \dot{\overline{\delta\boldsymbol{r}}} = \boldsymbol{0} \tag{4.29} $$

gefordert werden. Nach Ausklammern der überlagerten konstanten Krümmungsgeschwindigkeit $\boldsymbol{K}$ ergibt sich somit

$$ \boldsymbol{K} \cdot \Big\{ \int\limits_v \varrho \sum_{i=1}^{n} {}^i\alpha \, \delta\boldsymbol{r} \times ({}^i\boldsymbol{d} \times {}^i\dot{\boldsymbol{v}}) \, \mathrm{d}V - \int\limits_v \varrho \sum_{i=1}^{n} {}^i\alpha \delta\boldsymbol{r} \times ({}^i\boldsymbol{d} \times {}_i\boldsymbol{f}) \, \mathrm{d}V - \int\limits_o \sum_{i=1}^{n} \delta\boldsymbol{r} \times ({}^i\boldsymbol{d} \times {}_i\boldsymbol{s}) \, \mathrm{d}O \Big\} = 0 \, . \tag{4.30} $$

Ersetzt man in der verbleibenden Beziehung zunächst $\delta \boldsymbol{r}$ durch $\boldsymbol{r} - \boldsymbol{r}_0$, so kann bei beliebigem, aber konstantem Wert für $\boldsymbol{r}_0$ über

$$\boldsymbol{r}_0 \times \left\{ \int_v \varrho \sum_{i=1}^n {}^i\alpha({}^i\boldsymbol{d} \times {}^i\dot{\boldsymbol{v}})\, \mathrm{d}V - \int_v \varrho \sum_{i=1}^n {}^i\alpha({}^i\boldsymbol{d} \times {}_i\boldsymbol{f})\, \mathrm{d}V - \int_o \sum_{i=1}^n ({}^i\boldsymbol{d} \times {}_i\boldsymbol{s})\, \mathrm{d}O \right\} = \boldsymbol{0} \tag{4.31}$$

die Restgleichung

$$\int_v \boldsymbol{r} \times \varrho \sum_{i=1}^n {}^i\alpha({}^i\boldsymbol{d} \times {}^i\dot{\boldsymbol{v}})\, \mathrm{d}V - \int_v \boldsymbol{r} \times \varrho \sum_{i=1}^n {}^i\alpha({}^i\boldsymbol{d} \times {}_i\boldsymbol{f})\, \mathrm{d}V - \int_o \boldsymbol{r} \times \sum_{i=1}^n ({}^i\boldsymbol{d} \times {}_i\boldsymbol{s})\, \mathrm{d}O = \boldsymbol{0} \tag{4.32}$$

weiter reduziert werden. Nach Anwendung des Gaussschen Integralsatzes auf (4.32)

$$\int_o \left\{ \boldsymbol{r} \times \sum_{i=1}^n ({}^i\boldsymbol{d} \times {}_i\boldsymbol{s}) \right\} \mathrm{d}O = \int_v \left\{ \boldsymbol{r} \times \sum_{i=1}^n {}_i\hat{S}^{jk}\boldsymbol{g}_j \right\}_{,k} \mathrm{d}V \tag{4.33}$$

folgt bei Beachtung von (4.31) und Einführung eines Schnittkraftvektors ${}_i\hat{\boldsymbol{s}}$ und dazugehörigen Tensors ${}_i\hat{\mathbf{S}}$

$${}_i\hat{\boldsymbol{s}} = {}^i\boldsymbol{d} \times {}_i\boldsymbol{s}\,, \qquad {}_i\hat{\mathbf{S}} \cdot \boldsymbol{n} = {}_i\hat{\boldsymbol{s}} = ({}_i\hat{S}^{jk}\boldsymbol{g}_j\boldsymbol{g}_k) \cdot n_l\boldsymbol{g}^l = {}_i\hat{S}^{jl}n_l\boldsymbol{g}_j \tag{4.34 a, b}$$

mit

$$\boldsymbol{r} \times \left\{ \int_v \varrho \sum_{i=1}^n {}^i\alpha({}^i\boldsymbol{d} \times {}^i\dot{\boldsymbol{v}})\, \mathrm{d}V - \int_v \varrho \sum_{i=1}^n {}^i\alpha({}^i\boldsymbol{d} \times {}_i\boldsymbol{f})\, \mathrm{d}V - \int_v \left(\sum_{i=1}^n {}_i\hat{S}^{jk}|_k\, \boldsymbol{g}_j \right) \mathrm{d}V \right\} - \int_v \boldsymbol{r}_{,k} \times \sum_{i=1}^n {}_i\hat{S}^{jk}\boldsymbol{g}_j\, \mathrm{d}V = \boldsymbol{0}$$

die endgültige Symmetriebeziehung der Direktorschnittkräfte (4.35)

$$\sum_{i=1}^n \int_v \boldsymbol{g}_k \times {}_i\hat{S}^{jk}\boldsymbol{g}_j\, \mathrm{d}V = \boldsymbol{0}\,, \tag{4.36}$$

oder bei vorausgesetzter Stetigkeit der Lösung die lokale Form

$$\sum_{i=1}^n \boldsymbol{g}_k \times {}_i\hat{S}^{jk}\boldsymbol{g}_j = \boldsymbol{0}\,, \qquad \sum_{i=1}^n {}_i\hat{S}^{jk} = \sum_{i=1}^n {}_i\hat{S}^{kj}\,. \tag{4.37 a, b}$$

Ersetzt man in der Herleitung ${}_i\boldsymbol{s}$ durch den zugehörigen Spannungstensor ${}_i\mathbf{T}$, so läßt sich die Symmetriebeziehung auch in den Tensorkomponenten von ${}_i\mathbf{T}$ ausdrücken

$$\sum_{i=1}^n \varepsilon_{lm}{}^n \varepsilon_{jk}{}^m\, {}^id^j\, {}_iT^{kl}\boldsymbol{g}_n = \boldsymbol{0}\,, \tag{4.38}$$

wobei $\boldsymbol{\varepsilon}$ als Permutationstensor verwendet wurde. In diesem Zusammenhang sollte auf einen wesentlichen Unterschied zu den Arbeiten von Rivlin [13] und Stojanović [15] hingewiesen werden, von denen der erste dem Mikrokontinuum alle Eigenschaften eines Makrokontinuums zuordnet, also auch eine dem Boltzmannschen Axiom parallele Symmetrieforderung für den Mikrospannungstensor. Stojanović hingegen verlangt die Invarianz des Beanspruchungszustands eines Körpers bei Überlagerung einer „uniformen" Direktorbewegung. Beiden Arbeiten gemeinsam ist die Verwendung von Annahmen, die unmittelbar dem Mikrokontinuum zuzuordnen sind, während die hier vorgelegten Resultate erzielt werden konnten, indem die *Kinematik des Makrokontinuums erweitert* wurde. Dies hat jedoch zur Folge, daß im Gegensatz zu [13], [15] neben dem Kräfte- und Momentengleichgewicht nur *eine* vektorielle Zusatzbeziehung für die Summe der Direktorspannungen angebbar ist. Beschränkt man sich auf nur einen Direktor bzw. eine in sich starre Direktorgruppe im Sinne eines Cosseratkontinuums, so wird diese Einschränkung unbedeutend.

Nach der Ermittlung der Gleichgewichtsbeziehungen fehlt schließlich die Einführung geeigneter Verzerrungstensoren und die Herleitung entsprechender konstitutiver Gesetze. Zu einer ersten Formulierung für die innere Energie gelangt man nach Überführung des Energiesatzes gem. (2.1) in die lokale Form bei gleichzeitiger Beachtung der Beziehungen für das Kräftegleichgewicht und die Direktorrestbelastung

$$\varrho \dot{U} = \varrho r - h^k|_k + v_{l\|k}\, T^{lk} + \sum_{i=1}^n {}^iv_j\, {}_i\hat{f}^j + \sum_{i=1}^n {}^iv_{l\|k}\, {}_iT^{lk}\,. \tag{4.39}$$

Unschön ist in dieser Beziehung das Vorhandensein einer belastungsbezogenen Größe ${}_i\hat{\boldsymbol{f}}$, denn gemäß (4.26) entspricht sie gerade dem nicht im Gleichgewicht befindlichen Rest der Direktorbelastung. Wenn aber $\varrho \dot{U}$ unabhängig von einer möglichen Starrkörperbewegung sein soll, so muß eine Formulierung unter ausschließlicher Verwendung von Spannungs- und Verzerrungstensoren bzw. deren Geschwindigkeiten angebbar sein. Für die Art der Umschreibung liefert (4.27) einen Hinweis. Es ist jedoch zu beachten, daß die Anteile der Direktorrestbelastung in Richtung ihres jeweiligen Direktors aufgrund des Kreuzproduktes nicht in die Symmetriebedingung eingehen und von daher einer getrennten Betrachtung bedürfen. In der im folgenden Kapitel im einzelnen ausgeführten Herleitung für ein 1-Direktorkontinuum wird sich zeigen, daß die Forderung nach Gleichgewicht der Direktorschnittgrößen in Richtung des zugehörigen Direktors

$${}_i\hat{\boldsymbol{f}} \cdot {}^i\boldsymbol{d} = 0 \tag{4.40}$$

zu der gewünschten belastungsfreien Formulierung für die innere Energie führt.

Ohne vorherige Kenntnis der spezifischen Materialeigenschaften eines Direktorkontinuums wird man zunächst nur verlangen können, die Entropieproduktionsungleichung einzuhalten. Nach Einführung der freien Helmholtzschen Energie ϱA als

$$A = U - ST \tag{4.41}$$

ergibt sich wegen der Kopplung von (2.1) und (4.1) über die Wärmequelle r und den Wärmefluß h die Entropiebilanz zu

$$-\varrho\dot{A} - \varrho S\dot{T} - \frac{1}{T}h^k T_{,k} + T^{lk}v_{l\|k} + \sum_{i=1}^{n} {}_i\hat{f}^j\,{}^i v_j + \sum_{i=1}^{n} {}_iT^{lk}\,{}^i v_{l\|k} \geqq 0\,. \tag{4.42}$$

Zur Herleitung allgemeiner Materialgleichungen werden beispielsweise von ERINGEN [6], ODEN/REDDY [12] sogenannte Prinzipe bzw. a priori zu erfüllende Forderungen formuliert. In erweiterter Anwendung der bisher verwendeten Invarianz gegen Starrkörperbewegungen können zur Formulierung der freien HELMHOLTZschen Energie nur *objektive* Größen, d. h. Tensormaße für Spannungen und Verzerrungen herangezogen werden. Aus dem ursprünglichen Ansatz für ϱA

$$\varrho A = \varrho A(\boldsymbol{r}, \boldsymbol{r}_{,j}, {}^i\boldsymbol{d}, {}^i\boldsymbol{d}_{,j}, T, T_{,j}) \tag{4.43}$$

wird

$$\varrho A = \varrho A(\gamma_{lk}, {}^i\delta^j, {}^i\varkappa^j{}_k, T) \tag{4.44}$$

mit

$$\gamma_{lk} = \tfrac{1}{2}(g_{lk} - G_{lk})\,, \quad \dot{\gamma}_{lk} = (v_{l\|k} + v_{k\|l})\,, \tag{4.45 a, b}$$

$${}^i\delta^j = {}^id^j - {}^iD^j\,, \quad {}^i\dot{\delta}^j = {}^i\dot{d}^j\,, \tag{4.46 a, b}$$

$${}^i\varkappa^j{}_k = {}^ik^j{}_k - {}^iK^j{}_k\,, \quad {}^i\dot{\varkappa}^j{}_k = {}^i\dot{d}^j{}_{|k}\,. \tag{4.47 a, b}$$

Dabei sind g_{lk} die Metrik und ${}^ik^j{}_k$ die Krümmung des Direktorfeldes im aktuellen Zustand sowie G_{lk} und ${}^iK^j{}_k$ die entsprechenden Größen des Ausgangszustands.

Über

$$-\varrho\left\{\frac{\partial A}{\partial\gamma_{lk}}\dot{\gamma}_{lk} + \sum_{i=1}^{n}\left(\frac{\partial A}{\partial\,{}^i\delta^j}{}^i\dot{\delta}^j + \frac{\partial A}{\partial\,{}^i\varkappa^j{}_k}{}^i\dot{\varkappa}^j{}_k\right) + \frac{\partial A}{\partial T}\dot{T}\right\} -$$

$$-\varrho S\dot{T} - \frac{1}{T}h^kT_{,k} + T^{lk}v_{l\|k} + \sum_{i=1}^{n}{}_i\hat{f}^j\,{}^iv_j + \sum_{i=1}^{n}{}_iT^{lk}\,{}^iv_{l\|k} \geqq 0 \tag{4.48}$$

lassen sich bei Voraussetzung freier Variierbarkeit von $\boldsymbol{v}$ und ${}^i\boldsymbol{v}$ konstitutive Beziehungen für T^{lk}, ${}_i\hat{f}^j$, ${}_iT^{lk}$ und T angeben. Auf eine Wiedergabe wird an dieser Stelle zugunsten einer ausführlichen Erörterung einer numerisch verwertbaren Theorie eines einfachsten 1-Direktorkontinuums im folgenden Kapitel verzichtet.

5. Lineare Theorie einer Scheibe für transversalisotropes Material bei nur einem Direktor

Die Ausführungen dieses Abschnitts beziehen sich auf den Sonderfall einer „einfachsten" Direktortheorie. Darunter mögen folgende Voraussetzungen zusammengefaßt sein:

- Beschränkung auf einen Direktor,
- homogenes Direktorfeld,
- transversalisotropes, elastisches Material,
- kleine Verzerrungen bei kleinen Deformationen,
- Gleichgewicht am unverformten System.

Die Einarbeitung der oben genannten Vereinfachungen in die Feldgleichungen der in Kap. 4 bereitgestellten allgemeinen Beziehungen führt zu einer geometrisch und physikalisch linearen Theorie. Setzt man ferner statisches Gleichgewicht voraus, so erhält man neben der unveränderten Beziehung für den Massenerhalt die erste CAUCHYsche Bewegungsgleichung in der Form

$$\varrho f^k + T^{kj}|_j = 0\,. \tag{5.1}$$

Das Momentengleichgewicht vereinfacht sich über

$$\boldsymbol{D} \times {}_1\hat{\boldsymbol{f}} + \boldsymbol{G}_l \times T^{jl}\boldsymbol{G}_j + \boldsymbol{D}_{,k} \times {}_1T^{jk}\boldsymbol{G}_j = \boldsymbol{0} \tag{5.2}$$

aufgrund der Annahme eines homogenen, d. h. überall gleichgerichteten Einheitsdirektorfeldes in Richtung der Basis $\boldsymbol{e}_3$ mit

$$\boldsymbol{D}_{,l} = \boldsymbol{0} \tag{5.3}$$

zu

$$\boldsymbol{D} \times {}_1\hat{\boldsymbol{f}} + \boldsymbol{G}_l \times T^{jl}\boldsymbol{G}_j = \boldsymbol{0}\,. \tag{5.4}$$

Löst man diese Beziehung nach ${}_1\hat{\boldsymbol{f}}$ auf, so ergibt sich in Komponentenschreibweise

$${}_1\hat{f}^l = \alpha D^l + \frac{1}{D}D_i\{T^{il} - T^{li}\} \tag{5.5}$$

mit α als freiem Parameter. Eine weitere Beziehung für die Direktorrestbelastung liefert (4.26). Über die aus (4.38) gewonnenen vereinfachten Symmetriebedingungen des Direktorspannungstensors

$$D_i\,{}_1T^{kk} = D_k\,{}_1T^{ik} \tag{5.6}$$

gelangt man bei Beachtung von

$$D_\beta = 0 \tag{5.7}$$

mit $\beta = 1, 2$ zu

$$_1\hat{f}^\beta = {}_1T^{\beta\gamma}|_\gamma + \varrho\, {}^1\alpha\, {}_1f^\beta\,, \qquad {}_1\hat{f}^3 = {}_1T^{3i}|_i + \varrho\, {}^1\alpha\, {}_1f^3\,. \tag{5.8a, b}$$

Obwohl mit (5.6) eine Beziehung vorliegt, die eine weitgehende Einschränkung eines möglichen Direktorspannungszustands gestattet, gelingt es weder mit (5.5) noch mit (5.8), die Größe $_1\hat{f}^i$ vollständig zu bestimmen. Erst mit der Forderung, daß α verschwindet bzw. daß in Richtung des Direktors das Gleichgewicht allein durch $_1T^{jk}$ erfüllbar ist, lassen sich die Gleichgewichtsbedingungen vervollständigen:

$$_1\hat{f}^\beta = \frac{1}{D} D^i \varepsilon_{il}{}^\beta\, \varepsilon_{jk}{}^l\, T^{kj}\,, \qquad {}_1\hat{f}^3 = 0\,, \tag{5.9 a, b}$$

$$0 = {}_1T^{3i}|_i + \varrho\, {}^1\alpha\, {}_1f^3\,. \tag{5.10}$$

Dabei entspricht die Aussage von (5.9b) der allgemeinen Beziehung von (4.40). Mit (5.9) läßt sich die innere Energie für den Sonderfall des einfachsten Direktorkontinuums angeben zu

$$\varrho\dot{U} = \varrho r - h^k|_k + T^{kl}\left(v_{k|l} + \frac{D^i}{D}\, \varepsilon_{ij}{}^m\, \varepsilon_{lk}{}^j\, {}^1v_m\right) + {}_1T^{lk}\, {}^1v_{l|k}\,. \tag{5.11}$$

Aus der Entropieproduktionsungleichung (4.42) wird

$$-\varrho\dot{A} - \varrho S\dot{T} - \frac{1}{T} h^k T_{,k} + T^{kl}\lambda_{kl} + {}_1T^{kl}\varkappa_{kl} \geqq 0\,, \tag{5.12}$$

wobei in Anlehnung an Besdo [3] λ_{kl} und $\varkappa_{kl}$ als Tensoren der erweiterten Verzerrungsgeschwindigkeit und Verkrümmungsgeschwindigkeit eingeführt wurden.

Eine eingehende Analyse der Beziehung (5.12) hinsichtlich der Zweckmäßigkeit verschiedener Verzerrungsmaße für eine widerspruchsfreie Herleitung konstitutiver Gesetze führt in Abweichung von der Cosseratheorie zu dem mit (4.44) angegebenen Ansatz für die freie Helmholtzsche Energie. Ohne die z. T. langwierige Herleitung der Zeitableitungen im einzelnen angeben zu wollen, zerfällt die Entropieproduktionsungleichung in folgende konstitutive Beziehungen

$$T^{k\alpha} = \varrho \frac{\partial A}{\partial \gamma_{k\alpha}}, \qquad T^{\alpha 3} = \varrho \frac{\partial A}{\partial \gamma_{\alpha 3}} - \varrho \frac{\partial A}{\partial^1\delta^j} g^{j\alpha}\,, \qquad T^{33} = \varrho \frac{\partial A}{\partial \gamma_{33}}, \tag{5.13a, b, c}$$

$$\varrho \frac{\partial A}{\partial^1\delta^j} g^{j\alpha} = T^{3\alpha} - T^{\alpha 3}\,, \tag{5.14}$$

$$_1T^{kl} = \varrho \frac{\partial A}{\partial^1\varkappa^r{}_l} g^{kr}\,, \tag{5.15}$$

sowie entsprechende Gleichungen für den Temperaturfluß und die Entropie

$$-\frac{1}{T} h^k T_{,k} \geqq 0\,, \qquad S = -\frac{\partial A}{\partial T}. \tag{5.16), (5.17}$$

Die Annahme elastischen Materialverhaltens gestattet bei Vorgabe eines im Ausgangszustand spannungsfreien Kontinuums im Einklang mit Green/Zerna [8] die Darstellung der freien Helmholtzschen Energie als quadratisches Funktional der Verzerrungsmaße. Unter Verzicht auf den Einfluß der Temperatur gilt

$$\begin{aligned} 2\varrho_0 A = {} & E_1{}^{(kl)(ij)}\, \gamma_{(kl)}\, \gamma_{(ij)} + E_2{}^{(kl)}{}_m\, \gamma_{(kl)}\, {}^1\delta^m + E_3{}^{(kl)}{}_j{}^m\, \gamma_{(kl)}\, {}^1\varkappa^j{}_m + E_{4mn}\, {}^1\delta^m\, {}^1\delta^n + E_{5mj}{}^k\, {}^1\delta^m\, {}^1\varkappa^j{}_k + \\ & + E_{6j}{}^k{}_l{}^m\, {}^1\varkappa^j{}_k\, {}^1\varkappa^l{}_m\,. \end{aligned} \tag{5.18}$$

Die zugehörigen Spannungsverzerrungsbeziehungen lauten:

$$T^{k\alpha} = E_1{}^{(k\alpha)(ij)}\, \gamma_{(ij)} + \tfrac{1}{2} E_2{}^{(k\alpha)}{}_m\, {}^1\delta^m + \tfrac{1}{2} E_3{}^{(k\alpha)}{}_j{}^m\, {}^1\varkappa^j{}_m\,, \tag{5.19a}$$

$$T^{\alpha 3} = E_1{}^{(\alpha 3)(ij)}\, \gamma_{(ij)} + \tfrac{1}{2} E_2{}^{(\alpha 3)}{}_m\, {}^1\delta^m + \tfrac{1}{2} E_3{}^{(\alpha 3)}{}_j{}^m\, {}^1\varkappa^j{}_m - \tfrac{1}{2} G^{m\alpha}\{E_2{}^{(kl)}{}_m\, \gamma_{(kl)} + 2E_{4mn}\, {}^1\delta^n + E_{5mj}{}^k\, {}^1\varkappa^j{}_k\}\,, \tag{5.19b}$$

$$T^{33} = E_1{}^{33(ij)}\, \gamma_{(ij)} + \tfrac{1}{2} E_2{}^{33}{}_m\, {}^1\delta^m + \tfrac{1}{2} E_3{}^{33}{}_j{}^m\, {}^1\varkappa^j{}_m\,, \tag{5.19c}$$

$$T^{3\alpha} - T^{\alpha 3} = \tfrac{1}{2} G^{m\alpha}\{E_2{}^{(kl)}{}_m\, \gamma_{(kl)} + 2E_{4mn}\, {}^1\delta^n + E_{5mj}{}^k\, {}^1\varkappa^j{}_k\}\,, \tag{5.20}$$

$$_1T^{kl} = \tfrac{1}{2} G^{kr}\{E_3{}^{(ij)}{}_r{}^l\, \gamma_{(ij)} + E_{5mr}{}^l\, {}^1\delta^m + 2E_{6r}{}^l{}_i{}^j\, {}^1\varkappa^i{}_j\}\,. \tag{5.21}$$

In (5.18) bis (5.21) bezeichnet die runde Klammer einen in den einbezogenen Indizes symmetrischen Tensor und G^{jk} die Metrik im Referenzzustand. Die Zahl der unbekannten Materialkenngrößen läßt sich aus der Gesamtzahl der unabhängigen Verzerrungen, 6 für γ_{lk}, 3 für $^1\delta^m$ und zunächst 9 für $^1\varkappa^i{}_j$ zu 171 bestimmen. Insoweit läßt sich noch eine formale Analogie zur Cosseratheorie herstellen, für die Kessel [9] im allgemeinsten Fall ein ähnlich strukturiertes Materialgesetz bereitstellt.

Um zu einer praktisch handhabbaren Theorie zu gelangen, ist die Einarbeitung weiterer Vereinfachungen notwendig. Dabei stehen zwei Möglichkeiten offen. Eine erste besteht in der Annahme von Materialsymmetrien, von denen die transversale Isotropie, d. h. Umkehrbarkeit des Beanspruchungszustands bezüglich beliebiger Achsen und Rotation um eine fest vorgegebene Achse die höchstmögliche Symmetrievorgabe für ein unidirektionales Direktorkontinuum darstellt. Wie leicht einsehbar, kommt dabei für die Drehachse nur diejenige Richtung in Frage, die zur

Direktorrichtung parallel ist. Eine weitere Vereinfachungsmöglichkeit besteht in der Übertragung der durch (5.6) festgestellten Symmetrie des Direktorspannungstensors auf die Ermittlung eines entsprechenden Direktorverzerrungstensors, da bei sechs unabhängigen Spannungskomponenten nur ebensoviele Deformationsmaße voneinander unabhängig sein können. Analog zur klassischen Kontinuumsmechanik, die wegen der Symmetrie des Spannungstensors die Schubverzerrungen (z. B. $u_{1|2}$ und $u_{2|1}$) zu einer Summe zusammenfaßt — man vergleiche dazu die Ermittlung der Verschiebungsarbeit der Schubspannungen $\sigma^{21} = \sigma^{12}$

$$\sigma^{12}u_{1|2} + \sigma^{21}u_{2|1} = \sigma^{12}(u_{1|2} + u_{2|1}) = 2\sigma^{12}\varepsilon_{12}\,, \tag{5.22}$$

— kann aus der Beziehung für den Arbeitsanteil der Direktorspannungen ein symmetrisches Verzerrungsmaß konstruiert werden.

Im Rahmen einer linearen Theorie lassen sich die gemäß (4.45) bis (4.47) eingeführten Verzerrungsmaße aus dem unbekannten Verschiebungsfeld $\boldsymbol{u}$ und dem Direktorverschiebungsfeld ${}^1\boldsymbol{u}$ ohne weiteres bestimmen:

$$\gamma_{ik} = \tfrac{1}{2}\,(u_{i|k} + u_{k|i})\,, \tag{5.23}$$

$$\delta^i = {}^1u^i - u^j|_i\, D_j\,, \qquad i \neq j\,, \qquad {}^1\varkappa^j{}_k = {}^1u^j|_k\,. \qquad (5.24), (5.25)$$

Dabei ergibt sich der Ausdruck für δ^i als Ergebnis einer zusätzlichen Linearisierung, denn bei der in (4.46) angegebenen Differenz ist zu beachten, daß ${}^i d^j$ und ${}^i D^j$ auf verschiedene Basen bezogen sind.

Mit (5.25) folgt für den Arbeitsanteil der Direktorspannungen bei Verwendung von (5.7)

$${}_1T^{ij}\,{}^1\varkappa_{ij} = {}_1T^{\alpha\beta}\,{}^1u_{\alpha|\beta} + {}_1T^{3i}\,{}^1u_{3|i} \tag{5.26}$$

mit der Nebenbedingung

$${}_1T^{\alpha\alpha} = 0\,. \tag{5.27}$$

Aus (5.26) läßt sich in Verbindung mit (5.27) der ursprüngliche Verkrümmungstensor ${}^1\varkappa_{ij}$ durch

$$\varkappa_{11} = -{}^1u_{2|1}\,, \qquad \varkappa_{22} = {}^1u_{1|2}\,, \qquad \varkappa_{12} = \varkappa_{21} = \tfrac{1}{2}\,({}^1u_{1|1} - {}^1u_{2|2})\,, \qquad (5.28\text{ a, b, c})$$

$$\varkappa_i = {}^1u_{3|i} \tag{5.29}$$

ersetzen, wobei ${}_1T^{\alpha\beta}$ gleichzeitig durch $\hat{S}^{\alpha\beta}$ aus (4.33) ersetzt wird:

$$\hat{S}^{11} = -{}_1T^{21}\,, \qquad \hat{S}^{12} = {}_1T^{11} = \hat{S}^{21}\,, \qquad \hat{S}^{22} = {}_1T^{12}\,. \qquad (5.30\text{ a, b, c})$$

Führt man die neu definierten Verzerrungsmaße in den Ansatz für die freie HELMHOLTZsche Energie ein, so wird aus (5.18)

$$\begin{aligned} 2\varrho_0 A = {} & E_1{}^{(kl)(ij)}\,\gamma_{kl}\gamma_{ij} + E_2{}^{(kl)}{}_m\,\gamma_{kl}\,\delta^m + E_1{}^{(kl)(\alpha\beta)}\,\gamma_{kl}\,\varkappa_{(\alpha\beta)} + E_4{}^{(kl)i}\,\gamma_{kl}\,\varkappa_i + E_{5mn}\delta^m\delta^n + \\ & + E_{6m}{}^{(\alpha\beta)}\,\delta^m\,\varkappa_{(\alpha\beta)} + E_{7m}{}^i\,\delta^m\,\varkappa_i + E_8{}^{(\alpha\beta)(\gamma\delta)}\,\varkappa_{(\alpha\beta)}\,\varkappa_{(\gamma\delta)} + E_9{}^{(\alpha\beta)i}\,\varkappa_{(\alpha\beta)}\,\varkappa_i + E_{10}{}^{ij}\,\varkappa_i\varkappa_j\,, \end{aligned} \tag{5.31}$$

ein Materialgesetz mit immer noch 120 Unbekannten. Die Einarbeitung transversaler Isotropie führt schließlich zu dem einfachst möglichen, geschlossen herleitbaren Ausdruck für ϱA. Bild 7 zeigt die Anordnung der restlichen 13 Werkstoffkonstanten im Elastizitätstensor, wobei die Art der Darstellung in Anlehnung an GREEN/ZERNA [8] gewählt wurde.

Von besonderer Bedeutung ist in diesem Zusammenhang der durch die spezielle Wahl von $\boldsymbol{D}$ begründete Fortfall von δ^3, das wegen der in (5.11) deutlich werdenden Verknüpfung mit D^3 über den Permutationstensor keinen Beitrag zur inneren Energie liefern kann. Als Folge davon ergibt sich, wie bereits aus (5.13) ablesbar, eine Symmetrie des klassischen Spannungstensors bezüglich der Indexgruppe 12.

Um für die hier hergeleitete Direktortheorie auch den Nachweis der Anwendbarkeit führen zu können, sei abschließend die Formulierung eines quasi-ebenen Problems vorgestellt. Um alle bislang angegebenen Beziehungen

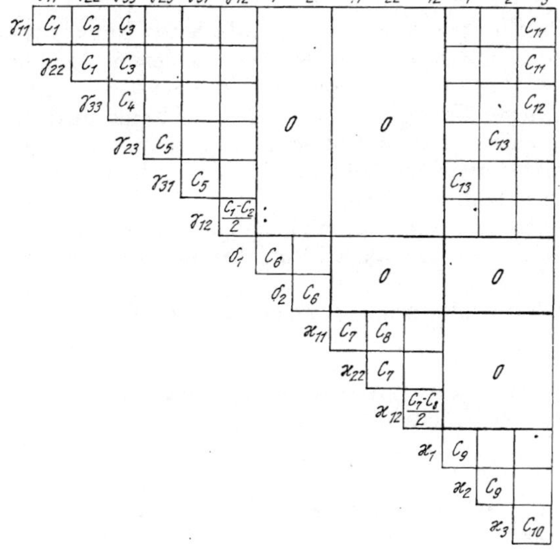

Bild 7. Elasitzitätskonstanten eines transversalisotropen 1-Direktorkontinuums

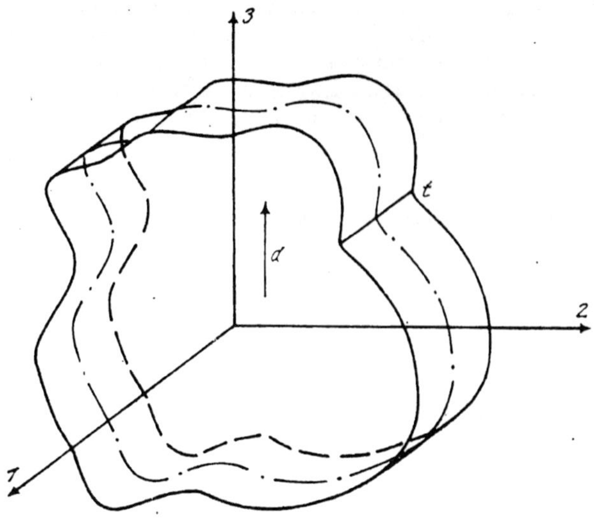

Bild 8. Geometrie einer Scheibe in der 2—3-Ebene

unverändert übernehmen zu können, werde angenommen, daß der betrachtete Kontinuumsausschnitt eine Scheibe kleiner Dicke t mit der 2—3-Ebene als Mittelfläche sei. Die Belastung sei konstant über die Scheibendicke und so gewählt, daß der resultierende Spannungs- und Verzerrungszustand symmetrisch zur Ebene $x_1 = 0$ vorausgesetzt werden kann.

Es besteht dann wegen der Unabhängigkeit der klassischen Kraftgrößen von einer möglichen Direktorbelastung neben dem ebenen Spannungszustand und dem ebenen Verzerrungszustand die Möglichkeit, zwei Mischformen vorauszusetzen. Die folgenden Ausführungen beziehen sich dementsprechend auf einen ebenen Kraftspannungszustand und einen ebenen Direktorverzerrungszustand. Danach gilt an der Oberfläche bei $\pm\, t/2$

$$T^{i1} = 0 \tag{5.32}$$

sowie im Körperinnern wegen (5.24), (5.28) und (5.29)

$${}^1u_1 = \delta^1 = 0\,. \tag{5.33}$$

Für die verwendeten Verzerrungsmaße ergibt sich bei Beachtung der konstitutiven Beziehungen

$$\gamma_{12} = \gamma_{13} = 0 \tag{5.34}$$

sowie

$$\varkappa_{11} = \varkappa_{22} = 0\,, \qquad \varkappa_{12} = -\tfrac{1}{2}\,{}^1u_{2|2}\,, \qquad \varkappa_1 = 0\,. \tag{5.35 a, b, c}$$

Da γ_{11} wegen vorausgesetzter Spannungsfreiheit in 1-Richtung keinen Beitrag zur inneren Energie liefert, kann es aus dem Ausdruck für die freie HELMHOLTZsche Energie eliminiert werden. Damit verbleiben schließlich 7 unbekannte Verzerrungsgrößen $\gamma_{22}, \gamma_{33}, \gamma_{23}, \delta_2, \varkappa_{12}, \varkappa_2$ und $\varkappa_3$ bzw. die entsprechenden 4 Komponenten der beiden unabhängigen Verschiebungsfelder $\boldsymbol{u}$ und ${}^1\boldsymbol{u}$.

Auf die Angabe entsprechend reduzierter Materialgleichungen wird verzichtet, da sie ohne Mühe durch Einarbeitung der Beziehungen (5.32) bis (5.35) erzeugt werden können.

6. Anmerkungen zum gewählten Lösungsverfahren

Für den praktischen Einsatz der in den vorangegangenen Kapiteln hergeleiteten einfachsten Direktortheorie fehlt am Ende nur noch ein leistungsfähiges Lösungsverfahren. Um sich nicht auf wenige geschlossen angebbare homogene Lösungen beschränken zu müssen, wird im folgenden die Anbindung an das Verfahren der Finiten Elemente aufgezeigt. Durch die Einführung der freien HELMHOLTZschen Energie als quadratisches Funktional in den Verzerrungen kann die Potentialeigenschaft der Verformungsenergie als gegeben vorausgesetzt werden. Gibt man sich mit einer Näherungslösung zufrieden, so besteht die Aufgabe darin, eine Reihe von Ansatzfunktion so zu verknüpfen, daß unter einer gegebenen äußeren Last ein Minimum an potentieller Energie erreicht wird. Beschränkt man z. B. Verschiebungsansätze auf Teilbereiche der Struktur und verbindet diese unter Einhaltung der Kontinuität zu einer im Gleichgewicht befindlichen Gesamtlösung, so ist dies nichts anderes als ein Vorgehen im Sinne eines weggrößenorientierten Finite-Element-Modells.

Ohne auf die Feinheiten in der Herleitung der für lineare Probleme verwendeten Formulierung

$$\mathbf{K}\cdot\boldsymbol{U} = \boldsymbol{R} \tag{6.1}$$

eingehen zu wollen, dies kann u. a. bei BATHE/WILSON [1] nachgelesen werden, seien im folgenden nur die direktorspezifischen Matrizen angegeben. Dabei errechnet sich die Steifigkeitsmatrix aus

$$\mathbf{K} = \int_V \boldsymbol{\Phi}^T\mathbf{B}^T\mathbf{C}\mathbf{B}\boldsymbol{\Phi}\,\mathrm{d}V\,. \tag{6.2}$$

In $\boldsymbol{U}$ sind die unbekannten Knotenverschiebungen und in $\boldsymbol{R}$ die vorgegebenen Knotenlasten zusammengefaßt. Während $\boldsymbol{\Phi}$ nur eine Interpolationsmatrix ist, die die diskreten Werte der Knotenverschiebungen in einen stetigen

Verlauf im Innern der Elemente verwandelt, also in keinem unmittelbaren Bezug zur Direktortheorie steht, beinhaltet **B** die Differentiationsvorschrift zur Bildung eines Verzerrungsvektors

$$\boldsymbol{\varepsilon} = (\gamma_{22}, \gamma_{33}, \gamma_{23}, \delta_2, \varkappa_{12}, \varkappa_2, \varkappa_3) \tag{6.3}$$

aus den Knotenverschiebungen

$$\boldsymbol{u} = (u_2, u_3, {}^1u_2, {}^1u_3) \,. \tag{6.4}$$

Die Matrix **C** beinhaltet das Materialgesetz zur Bestimmung des Spannungsvektors

$$\sigma = (T^{22}, T^{33}, T^{32}, T^{32} - T^{23}, {}_1T^{11}, {}_1T^{32}, {}_1T^{33}) \,, \tag{6.5}$$

aus dem zuvor ermittelten Verzerrungsvektor $\boldsymbol{\varepsilon}$.

Abschließend sei nur noch erwähnt, daß **B** und **C** aus (5.23), (5.24), (5.28) und (5.29) sowie (5.31) unmittelbar folgen, wenn diese in eine Matrizenschreibweise überführt werden.

7. Ergebnisse für eine schubbelastete Scheibe

Zum Abschluß möge anhand numerischer Ergebnisse belegt werden, daß die dargelegte Theorie geeignet ist, Effekte zu beschreiben, die im Rahmen einer klassischen Kontinuumstheorie nicht erfaßt werden können. Außerdem wird deutlich werden, daß auch die einfachste Direktortheorie bereits Phänomene einschließt, die ebenfalls über den Geltungsbereich der seit langem bekannten COSSERATtheorie hinausgehen und nur durch den Widerstand eines Direktorkontinuums gegen die Strukturkrümmung erklärbar sind.

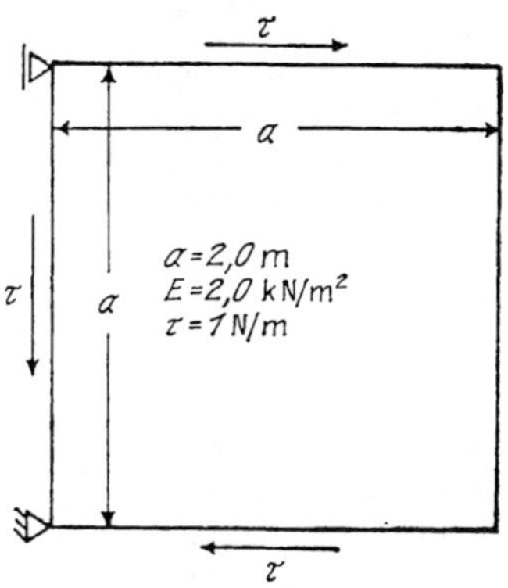

Bild 9. Geometrie und Belastung für ein Berechnungsbeispiel

Zu diesem Zweck seien die Ergebnisse für ein möglichst einfaches Beispiel, nämlich das einer an zwei Punkten gelagerten Scheibe unter konstanter Schubbelastung, angegeben. Um die Direktoreigenschaft auch an der Strukturdeformation deutlich werden zu lassen, sei am linken Rand eine Direktorverschiebung vollständig behindert.

Für die Beurteilung der Lösungen sind die Maßeinheiten der Zahlenwerte für Elastizitätskonstanten und die Belastung ohne Bedeutung, weshalb im folgenden nur noch das gegenseitige Verhältnis der Steifigkeiten angegeben wird.

Verzichtet man von vornherein auf eine Rückkopplung zwischen der Direktorlängsverformung $\varkappa_i$ und der Standardverzerrung γ_{ik}, so können die Elastizitätskonstanten $C_{10} - C_{13}$ mit Null angenommen werden. Die übrigen Materialparameter lassen sich entsprechend den beteiligten Verzerrungsmaßen bestimmten Grundverformungsmustern zuordnen:

C_1 bis C_5 der klassischen Kontinuumsverformung (γ_{ik}),
C_6, C_7 der Direktordrehung, Verkrümmung $(\delta_2, \varkappa_{12})$ und
C_8, C_9 der Längenänderung des Direktors $(\varkappa_2, \varkappa_3)$.

Damit sind C_6 und C_7 einer Materialeigenschaft zugeordnet, wie sie sich in ähnlicher Weise auch bei einem COSSERATkontinuum einstellen würde, während C_8 und C_9 einer unverwechselbaren Direktoreigenschaft zugerechnet werden müssen.

Die folgenden zwei Bilder mögen einen Eindruck davon vermitteln, welche Vielfalt von Lösungen im Rahmen einer Direktortheorie durch geeignete Wahl der Materialparameter eröffnet wird. Die Computerplots geben jeweils die Standardverformung und die zugehörige Direktorverformung wieder, wobei infolge verschwindender 1-Verschiebung die Komponenten in 2-Richtung horizontal und die Komponenten in 3-Richtung vertikal aufgetragen wurden.

Die Bilder 10 a, b gehören zu einer Materialkonfiguration, bei der C_1 bis C_5 wie für ein isotropes Material ohne Querdehnung vorgegeben wurden. Die Konstanten C_6, C_7 wurden dazu in gleicher Größe wie der Schubmodul eingeführt. In der zweiten Bildgruppe 11 a, b wurden zusätzlich C_8 und C_9 von praktisch Null ebenfalls auf den gleichen Wert wie C_6 angehoben. Gerechnet wurden diese Ergebnisse auf der Grundlage der vorgestellten linearen Theorie mit einem Raster von 8×8 isoparametrischen 8-Knotenelementen.

Ausgewählt wurden die Beispiele wegen der Möglichkeit einer eindeutigen Zuordnung zu bestimmten Deformationsmustern. Bild 10a zeigt in völliger Übereinstimmung mit der COSSERATtheorie eine ausschließlich lokale Wirkung der gewählten Direktorrandbedingung. Die Abweichung vom klassischen Ergebnis einer konstanten

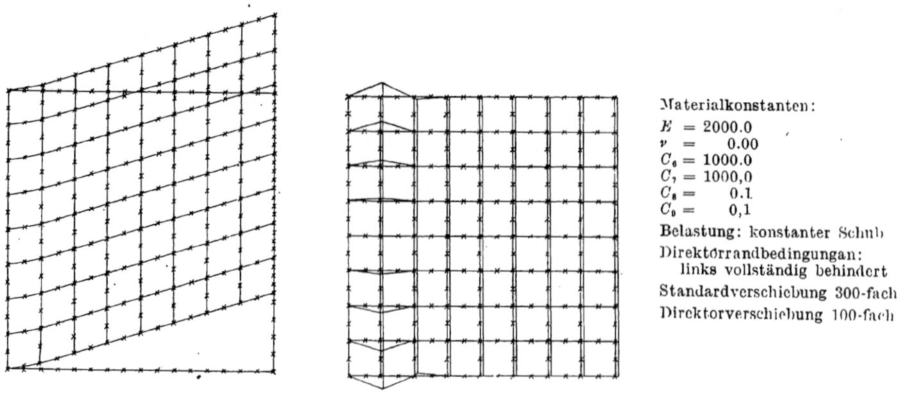

Bild 10. Computerplot einer Cosseratähnlichen Lösung
a) Standardverschiebung b) Direktorverschiebung

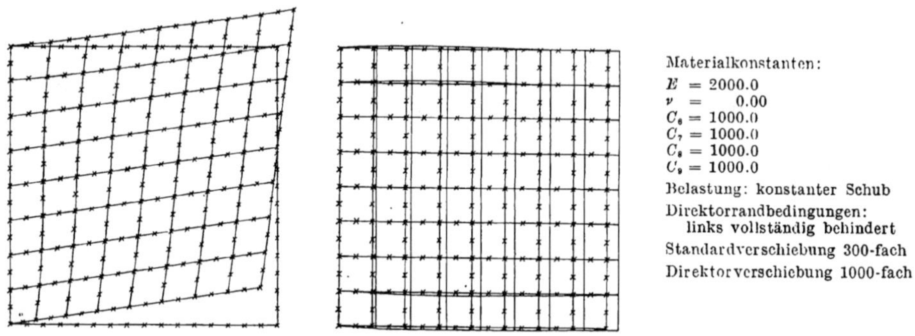

Bild 11. Computerplot einer „vollständigen" Direktorlösung
a) Standardverschiebung b) Direktorverschiebung

Schubverzerrung beschränkt sich auf die unmittelbare Nachbarschaft des gelagerten Randes. Dieses Ergebnis spiegelt sich auch im Direktordeformationsbild wider, das ebenfalls nur in Randnähe deutliche Deformationen aufweist. Hinzuzufügen ist in diesem Zusammenhang jedoch, daß eine weitere Erhöhung der Elastizitätskonstanten C_6 und C_7 ohne wesentlichen Einfluß auf das Deformationsbild bleibt.

Ganz anders verhält sich demgegenüber ein Kontinuum, bei dem durch Vorgabe von C_8 und C_9 die Steifigkeit eines Direktors in Längsrichtung in das Tragvermögen einer Struktur einbezogen wird. Die in gleichem Maßstab dargestellte Standardverschiebung offenbart gleich in zweifacher Hinsicht eine deutliche Abweichung von der klassischen Lösung. Neben einer erheblichen Reduzierung der Tangentialverschiebung auf ca. 45% des Ausgangswertes ist eine zusätzliche Deformation quer zur Schubrichtung zu beobachten. Der Einfluß der Direktorsteifigkeit ist auch in Bild 11b deutlich zu erkennen, in dem die Verschiebungskomponenten ${}^1u_2$ und ${}^1u_3$ im Vergleich zu Bild 10b mit zehnfacher Überhöhung dargestellt sind. Wie zu erwarten, hat sich die Spitze der ${}^1u_3$ Verschiebung, d. h. des Deformationsanteils in Richtung des Direktors, in der Nähe des festgehaltenen Randes abgebaut. Statt dessen stellt sich im gesamten Lösungsgebiet eine kontinuierlich veränderliche Direktordeformation ein. Anschaulich gesprochen heißt das, daß die Vorgabe der vollständigen Direktorsteifigkeit zur Einbeziehung des Widerstands eines Körpers gegen eine Orientierungsänderung bzw. Änderung seiner lokalen Krümmung in den global erkennbaren Tragmechanismus führt.

Im Gegensatz zur Cosserattheorie ist also festzustellen, daß die Annahme von Direktoren zu einer Körpereigenschaft führt, die nicht nur in der Art eines Randschichteffekts Einfluß auf die Spannungsverteilung und das Deformationsbild einer Struktur nimmt, sondern das Tragverhalten im ganzen erkennbar ändert.

Dieser Nachweis konnte im übrigen bei einer Vielzahl weiterer untersuchter Probleme ebenfalls geführt werden und läßt sich anhand einer Parameterstudie der Elastizitätskonstanten eindrucksvoll belegen. Auf Ergebnisse hierzu wird in diesem Zusammenhang verzichtet, da das Hauptanliegen dieser Arbeit darin bestand, das theoretische Rüstzeug vorzugeben und dessen Anwendbarkeit lediglich aufzuzeigen.

8. Schlußbemerkungen

Es war das erklärte Ziel dieser Arbeit, die rechnerische Behandlung orientierter Kontinua in ein neuartiges und dennoch praktikables theoretisches Konzept einzubetten. Mit den vorgelegten Ergebnissen darf dieser Zweck sicher als erfüllt angesehen werden. Darüber hinaus eröffnet sich aus der Verbindung von Direktortheorie und Finiter-Element-Methode erstmalig die Möglichkeit, die Hypothese von der Beschreibbarkeit orientierter Medien als ein Direktorkontinuum durch die Gegenüberstellung von experimentellen Ergebnissen und Rechnungen zu überprüfen. Dies gilt auch für komplexe Versuchsanordnungen, für die die Bereitstellung geschlossener Lösungen in der Regel auszuschließen ist.

In der Literatur (vgl. z. B. BECKER/HAUGER [2]) wird häufig der Versuch unternommen, die Relevanz entsprechender, von der klassischen BOLTZMANNschen Kontinuumsmechanik abweichender Theorien an Fragestellungen aus den Bereichen Fließvorgänge bei Metallen und Bestimmung mechanischer Eigenschaften von Granulaten bzw. hochpolymeren Kunststoffen aufzuzeigen. Es ist nicht von der Hand zu weisen, daß gerade bei Materialien mit ausgeprägter Feinstruktur Orientierungseffekte von größerer Bedeutung zu erwarten sind. Daraus läßt sich als nächstliegendes Forschungsziel die Notwendigkeit ableiten, ausgehend von den allgemeingültigen Beziehungen in Kap. 4, geometrische und physikalische Nichtlinearitäten in den Lösungsalgorithmus einzubeziehen, um z. B. das viskoelastische Fließen von Kunststoffen berücksichtigen zu können. Nicht übersehen werden sollte dabei allerdings, daß es bereits mit der bereitgestellten linearen Theorie möglich ist, grundsätzliche Aussagen zum Direktortragverhalten orientierter Kontinua zu treffen.

Literatur

1 BATHE, K.-J.; WILSON, E. L., Numerical Methods in Finite Element Analysis, Prentice Hall, Inc., Englewood Cliffs, New Jersey 1976.

2 BECKER, M.; HAUGER, W., Granular Material — an Experimental Realization of a Plastic Cosserat Continuum, Proceedings of the Int. Symp. on Mechanics of Inelastic Media and Structures, Warschau 1978.

3 BESDO, B., Ein Beitrag zur nichtlinearen Theorie des Cosserat-Kontinuums, Acta Mechanica, Vol. **20** (1974), S. 105—131.

4 BUGGISCH, H., Herleitung der mechanischen Bilanzgleichungen des Cosserat-Kontinuums aus der Energiegleichung und ihrem Verhalten beim Übergang zu rotierenden Systemen, ZAMM **53** (1973), S. T 68—T 69.

5 ERINGEN, A. C., Theory of Micropolar Elasticity. In: Liebowitz, H., Fracture, Vol. II. Academic Press, New York 1968, pp. 621 to 729.

6 ERINGEN, A. C., Mechanics of Continua, John Wiley & Sons, Inc., New York/London/Sydney 1967.

7 GREEN, A. E.; RIVLIN, R. S., The Relation between Director and Multipolar Theories in Continuum Mechanics, ZAMP **18** (1967), S. 208—218.

8 GREEN, A. E.; ZERNA, W., Theoretical Elasticity, 2nd Edition, At the Clarendon Press, Oxford 1968.

9 KESSEL, S., Lineare Elastizitätstheorie des anisotropen Cosserat-Kontinuums, Abh. der Braunschw. Wiss. Gesellschaft, Bd. XIV (1964), S. 1—22.

10 MINDLIN, R. D., Micro-Structure in Linear Elasticity, Arch. Rat. Mech. Anal., Vol. **16** (1964), pp. 51—78.

11 NAGHDI, P. M., The Theory of Shells and Plates. In: Flügge, S., Handbuch der Physk, Bd. VIa/2, Springer Verlag, Berlin/Heidelberg/New York 1972, S. 425—640.

12 ODEN, J. T.; REDDY, J. N., Variational Methods in Theoretical Mechanics, Springer Verlag, Berlin 1976.

13 RIVLIN, R. S., The Formulation of Theories in Generalized Continuum Mechanics and their Physical Significance, Symposia in Mathematica, Vol. **1** (1969), pp. 357—373.

14 ROTHERT, H., Direkte Theorie von Linien- und Flächentragwerken bei viskoelastischem Werkstoffverhalten, Technisch-wissenschaftliche Mitteilungen Nr. 73-2, IKIB, Ruhr-Universität Bochum 1973.

15 STOJANOVIĆ, R., On the Mechanics of Materials with Microstructure, Acta Mechanica, Vol. **15** (1972), S. 261—273.

16 ZASTRAU, B., Ein Beitrag zur Erweiterung klassischer Invarianzanforderungen für die Herleitung einer Direktortheorie, ZAMM **61** (1981), S. T 135—T 137.

Eingereicht am 15. 12. 80

Anschriften: Dr.-Ing. BERND ZASTRAU, Institut für Mechanik, Hochschule der Bundeswehr Hamburg, Holstenhofweg 85, D-2000 Hamburg 70, BRD;
Prof. Dr.-Ing. HEINRICH ROTHERT, Direktor des Instituts für Statik, Universität Hannover (TH), Callinstr. 32, D-3000 Hannover 1, BRD

ZAMM 61, 583 – 587 (1981)

K. N. SRIVASTAVA / O. P. GUPTA / R. M. PALAIYA

Interaction of Elastic Waves with a Griffith Crack Situated in an Infinitely Long Strip

Das Problem ist die Bestimmung der Spannungsverteilung in der Umgebung eines Griffith-Risses, der sich in einem unendlich langen elastischen Streifen befindet. Der Riß wird durch longitudinale elastische Wellen geöffnet, die normal zum Riß einfallen. Die Anwendung von Fourier-Transformationen reduziert das Problem auf eine Fredholmsche Integralgleichung zweiter Art. Die numerische Lösung dieser Gleichung wird zur Berechnung der Werte des Spannungskonzentrationsfaktors am Rand des Risses benutzt. Die Ergebnisse werden grafisch illustriert.

The problem is the determination of the distribution of stress in the neighbourhood of a Griffith crack situated in an infinitely long elastic strip. The crack is opened by longitudinal elastic waves, incident normally on the crack. Fourier transforms are used to reduce the problem to a Fredholm integral equation of the second kind. Numerical solution of this equation is used for calculating the values of dynamical stress intensity factor at the edge of the crack. The results are illustrated graphically.

Проблема состоит в определении распределения напряжений в окрестности трещины Гриффиса, которая расположена в безконечно длинной упругой полосе. Трещина расширяется за счёт продольных упругих волн, которые впадают нормально к трещине. Проблема сводится к решению интегрального уравнения Фредгольма второго рода с помощью преобразований Фурье. Численное решение этого уравнения используется для определения значений фактора концентрировании напражения у края трещины. Результаты изображены в виде графиков.

1. Introduction

In [1], we have studied the problem of finding the distribution of stresses in the neighbourhood of a Griffith crack located at the interface of two bonded dissimilar elastic half-planes. The crack is interacted by longitudinal elastic waves incident on its surface. In this paper we extend the method given in [1] to study a Griffith crack situated in an infinitely long elastic strip and interacted by incident longitudinal elastic waves. With the help of Fourier sine and cosine transforms, solutions of the displacement equations have been obtained, and the problem is reduced to that of solving a Fredholm integral equation of the second kind. This equation is solved numerically, numerical values of the dynamic stress intensity factor are calculated and shown graphically.

The study has been motivated by two definite objectives. First, to study the effect of the dynamic loading on the stress and displacement field and to compare the results for the limiting case when the wave frequency tends to zero with those obtained by SNEDDON and SRIVASTAV [3]. Secondly it is desired to study the effect of a finite boundary on the stress and displacement field due to dynamic loading and to compare the results with those derived by MAL [4].

2. Formulation of the Problem

Consider an infinitely long elastic strip of width $2h_1$ containing a Griffith crack located at $0 \leqq |x_1| \leqq a$, $x_2 = 0$. The crack is assumed to be excited by a normally incident longitudinal wave moving along the positive direction of the x_2 axis.

It is convenient to make all the quantities dimensionless by writing

$$\frac{x_1}{a} = x, \qquad \frac{x_2}{a} = y, \qquad \frac{h_1}{a} = h.$$

Thus the problem is to find the stress distribution near the crack edges subject to the following boundary conditions:

$$\sigma_{yy}(x, 0) = -p_s - p(x) \exp(-i\omega t), \qquad |x| < 1, \tag{2.1}$$

$$u_y(x, 0) = 0, \qquad 1 \leqq |x| \leqq h, \qquad \sigma_{xy}(x, 0) = 0, \qquad |x| \leqq h. \qquad (2.2), (2.3)$$

The edges of the strip are stress free, so that

$$\sigma_{xy}(\pm h, y) = 0 = \sigma_{xx}(\pm h, y), \qquad y \geqq 0. \tag{2.4}$$

Without any loss of generality, the boundary conditions (2.4) may be expressed as follows

$$F_s[\sigma_{xy}(\pm h, y); y \to \xi] = 0 = F_c[\sigma_{xx}(\pm h, y); y \to \xi] \tag{2.5}$$

where F_s and F_c denote Fourier sine and cosine transform respectively.

In condition (2.1) the static pressure p_s is assumed to be sufficiently large to ensure that the two faces of the crack do not come in contact during vibrations. Since the solution of the static problem may be superimposed on the dynamical problem, the condition (2.1) may be expressed as

$$\sigma_{yy}(x, 0) = -p(x) \exp(-i\omega t). \tag{2.6}$$

In what follows the time dependence of all the quantities assumed to be of the form $\exp(-i\omega t)$ will be suppressed.

In cartesian coordinates system the displacement vector $\boldsymbol{U}$ of a symmetrical problem has the form $(u_x, u_y, 0)$ and in the above circumstances u_x, u_y have the following form:

$$u_x = F_s[\{\exp(-\alpha y) - 2\alpha\beta(2\zeta^2 - k_2^2)^{-1}\exp(-\beta y)\}\, A(\zeta); \zeta \to x] + \\ + F_c[\{B(\xi)\sinh(\alpha_1 x) + C(\xi)\sinh(\beta_1 x)\}; \xi \to y], \tag{2.7}$$

$$u_y = F_c[\{\exp(-\alpha y) - 2\zeta^2(2\zeta^2 - k_2^2)^{-1}\exp(-\beta y)\}\, \alpha\zeta^{-1}A(\zeta); \zeta \to x] - \\ - F_s[\xi\alpha_1^2 B(\xi)\cosh(\alpha_1 x) + \beta_1\xi^{-1}C(\xi)\cosh(\beta_1 x)\}; \xi \to y]. \tag{2.8}$$

The corresponding expressions for the stress field are:

$$\mu^{-1}\sigma_{xy} = -F_s[\{\exp(-\alpha y) - \exp(-\beta y)\}\, 2\alpha A(\zeta); \zeta \to x] - \\ - F_s[\{2\xi B(\xi)\sinh(\alpha_1 x) + \xi^{-1}(2\xi^2 - k_2^2)\, C(\xi)\sinh(\beta_1 x)\}; \xi \to y], \tag{2.9}$$

$$\mu^{-1}\sigma_{yy} = -F_c[\{(2\zeta^2 - k_2^2)\exp(-\alpha y) - 4\alpha\beta\zeta^2(2\zeta^2 - k_2^2)^{-1}\exp(-\beta y)\}\, \zeta^{-1}A(\zeta); \zeta \to x] - \\ - F_c[\{\alpha_1^{-1}(2\alpha_1^2 + k_2^2)\, B(\xi)\cosh(\alpha_1 x) + 2\beta_1 C(\xi)\cosh(\beta_1 x)\}; \xi \to y], \tag{2.10}$$

$$\mu^{-1}\sigma_{xx} = F_c[\{(2\alpha^2 + k_2^2)\exp(-\alpha y) - 4\alpha\beta\zeta^2(2\zeta^2 - k_2^2)^{-1}\exp(-\beta y)\}\, \alpha^{-1}A(\zeta); \zeta \to x] + \\ + F_c[\{\alpha_1^{-1}(2\alpha_1^2 - k_2^2)\, B(\xi)\cosh(\alpha_1 x) + 2\beta C(\xi)\cosh(\beta_1 x)\}; \xi \to y] \tag{2.11}$$

where

$$\alpha = \begin{cases}(\zeta^2 - k_1^2)^{1/2}, & \zeta > k_1\\ -i(k_1^2 - \zeta^2)^{1/2}, & \zeta < k_1\end{cases}, \qquad \alpha_1 = \begin{cases}(\xi^2 - k_1^2)^{1/2}, & \xi > k_1\\ -i(k_1^2 - \xi^2)^{1/2}, & \xi < k_1\end{cases},$$

$$\beta = \begin{cases}(\zeta^2 - k_2^2)^{1/2}, & \zeta > k_2\\ -i(k_2^2 - \zeta^2)^{1/2}, & \zeta < k_2\end{cases}, \qquad \beta_1 = \begin{cases}(\xi^2 - k_2^2)^{1/2}, & \xi > k_2\\ -i(k_2^2 - \xi^2)^{1/2}, & \xi < k_2\end{cases},$$

$$k_2 = \left(\frac{\varrho}{\mu}\right)^{1/2}\omega a, \quad k_1 = \left(\frac{\varrho}{\lambda + 2\mu}\right)^{1/2}, \quad \omega a = k_2\frac{(1 - 2\eta)}{(2 - 2\eta)}, \tag{2.12}$$

λ and μ are Lamé's constants and ϱ denotes the density of the elastic material. The k_2 and k_1 are dimensionless wave numbers.

3. Solution of the Problem

It is clear from the expression (2.9) that the boundary condition (2.3) is automatically satisfied. The conditions (2.2) and (2.6) lead to the following set of dual integral equations:

$$\int_0^\infty \zeta^{-1}[1 + H(\zeta)]\,\varphi(\zeta)\sin(\zeta x)\,d\zeta \doteq Q(x), \qquad 0 \leqq |x| \leqq 1, \tag{3.1}$$

$$\int_0^\infty \zeta^{-1}\varphi(\zeta)\cos(\zeta_1 x)\,d\zeta = 0, \qquad 1 \leqq |x| \leqq h, \tag{3.2}$$

where

$$\varphi(\zeta) = \frac{2\alpha A(\zeta)\,(k_1^2 - k_2^2)}{2\zeta - k_2^2}, \qquad H(\zeta) = \frac{(2\zeta^2 - k_2^2)^2 - 4\alpha\beta\zeta^2}{2\alpha\zeta(k_1^2 - k_2^2)} - 1, \quad H(\zeta) \to 0, \quad \zeta \to \infty, \tag{3.3), (3.4}$$

$$Q(x) = \sqrt{\left(\frac{\pi}{2}\right)}\mu^{-1}\int_0^x p(x)\,dx - \int_0^\infty [\alpha_1^{-2}(2\alpha_1^2 + k_2^2)\sinh(\alpha_1 x)\, B(\xi) + 2\sinh(\beta_1 x)\, C(\xi)]\,d\xi. \tag{3.5}$$

To reduce the above dual integral equations to a single Fredholm integral equation we choose (see SNEDDON [5])

$$\varphi(\zeta) = \sqrt{\left(\frac{\pi}{2}\right)}\zeta\int_0^1 tg(t)\, J_0(\zeta t)\,dt. \tag{3.6}$$

Making use of (3.6) with equations (3.1) and (3.2) and using the results given in Appendix we get

$$g(t) + \int_0^1 ug(u)\, L_1(t, u)\,du = R(t) \tag{3.7}$$

where

$$L_1(t, u) = \int_0^\infty \zeta H(\zeta)\, J_0(\zeta t)\, J_0(\zeta u)\,d\zeta, \tag{3.8}$$

$$R(t) = \frac{2}{\mu\pi t}\frac{d}{dt}\int_0^t p(z)\,(t^2 - z^2)^{1/2}\,dz - \left(\frac{2}{\pi}\right)^{1/2}\int_0^\infty \left[\frac{2\alpha_1^2 + k_2^2}{\alpha_1}\, I_0(\alpha_1 t)\, B(\xi) + 2\beta_1 I_0(\beta_1 t)\, C(\xi)\right]d\xi. \tag{3.9}$$

For finding the unknown constants $B(\xi)$ and $C(\xi)$ we use the boundary conditions (2.5) which lead to the following relations:

$$2\xi^2 \sinh(\alpha_1 h) B(\xi) + (2\xi^2 - k_2^2) \sinh(\beta_1 h) C(\xi) = i_1(\xi) , \tag{3.10}$$

$$(2\xi^2 - k_2^2) \cosh(\alpha_1 h) B(\xi) + 2\alpha_1\beta_1 \cosh(B_1 h) C(\xi) = i_2(\xi) , \tag{3.11}$$

where

$$i_1(\xi) = -\frac{4\xi^2}{\pi}\int_0^\infty \frac{\alpha(k_1^2 - k_2^2) A(\zeta) \sin(\zeta h)}{\zeta(\alpha^2 + \xi^2)(\beta^2 + \xi^2)} d\zeta ,$$

$$i_2(\xi) = -\frac{2\alpha_1}{\pi}\int_0^\infty \left[\frac{\alpha(2\alpha^2 + k_2^2)}{\alpha^2 + \xi^2} - \frac{4\alpha\beta^2\zeta^2}{(2\zeta^2 - k_2^2)(\beta^2 - + \xi^2)}\right] A(\zeta) \cos(\zeta h) d\xi .$$

By solving the equations (3.10) and (3.11) and making use of equation (3.3), (3.6) and using the results given in Appendix we get

$$M(\xi) B(\xi) = -2\alpha_1\beta_1 \cosh(\beta_1 h) i_1(\xi) + (2\xi^2 - k_2^2) \sinh(\beta_1 h) i_2(\xi) , \tag{3.12}$$

$$M(\xi) C(\xi) = (2\xi^2 - k_2^2) \cosh(\alpha_1 h) i_1(\xi) - 2\xi^2 \sinh(\alpha_1 h) i_2(\xi) \tag{3.13}$$

where

$$M(\xi) = (2\xi^2 - k_2^2)^2 \sinh(\beta_1 h) \cosh(\alpha_1 h) - 4\alpha_1\beta_1\xi^2 \sinh(\alpha_1 h) \cosh(\beta_1 h) ,$$

$$i_1(\xi) = -\sqrt{\left(\frac{\pi}{2}\right)}\frac{\xi^2}{k_1^2 - k_2^2}\int_0^1 [(k_2^2 + 2\beta_1^2) \exp(-\beta_1 h) I_0(\beta_1 u) - (k_2^2 + 2\alpha_1^2) \exp(-\alpha_1 h) I_0(\alpha_1 h)] u g(u) du ,$$

$$i_2(\xi) = -\sqrt{\left(\frac{\pi}{2}\right)}\frac{1}{2(k_1^2 - k_2^2)}\int_0^1 [(2\alpha_1^2 + k_2^2)(2\beta_1^2 + k_2^2) \exp(-\alpha_1 h) I_0(\alpha_1 h) -$$

$$- 4\alpha_1\beta_1(\beta_1^2 + k_2^2) \exp(-\beta_1 h) I_0(\beta_1 u)] u g(u) du .$$

Substituting these values of $B(\xi)$ and $C(\xi)$ in (3.9) we obtain

$$R(t) = \frac{2}{\mu\pi t}\frac{d}{dt}\int_0^t p(z)(t^2 - z^2)^{1/2} dz + \int_0^1 u[L_2(t, u) + L_3(t, u)] g(u) du \tag{3.14}$$

where

$$L_2(t, u) = -\int_0^\infty [\{(2\alpha_1^2 + k_2^2) I_0(\alpha_1 t) \cosh(\beta_1 h) - (2\beta_1^2 + k_2^2) I_0(\beta_1 t) \cosh(\alpha_1 t) \times$$

$$\times \{(2\alpha_1^2 + k_2^2) \exp(-\alpha_1 h) I_0(\alpha_1 u) - (2\beta_1^2 + k_2^2) \exp(-\beta_1 h) I_0(\beta_1 u)\}] \frac{2\beta_1\xi^2 d\xi}{M(\xi)(k_1^2 - k_2^2)},$$

$$L_3(t, u) = -\int_0^\infty [\{4\alpha_1\beta_1 I_0(\beta_1 t) \sinh(\alpha_1 h) - (2\alpha_1^2 + k_2^2)(2\beta_1^2 + k_2^2) I_0(\alpha_1 t) \sinh(\beta_1 t)\} \times$$

$$\times \{4\alpha_1\beta_1 \exp(-\beta_1 h) I_0(\beta_1 u) - (2\alpha_1^2 + k_2^2)(2\beta_1^2 + k_2^2) \exp(-\alpha_1 h) I_0(\alpha_1 u)\}] \frac{d\xi}{\alpha M(\xi)(k_1^2 - k_2^2)} .$$

Making use of expression (3.14) with (3.7), we find

$$g(t) + \int_0^1 u g(u) L(t, u) du = j(t) \tag{3.15}$$

where

$$L(t, u) = L_1(t, u) + L_2(t, u) + L_3(t, u), \qquad j(t) = \frac{2}{\mu\pi t}\frac{d}{dt}\int_0^t p(z)(t^2 - z^2)^{1/2} dz .$$

For the sake of simplicity we take $p(x) = p_0$ (say constant) and assume $g(t) = \mu^{-1}p_0 g_1(t)$, then the equation (3.15) simplifies to

$$g_1(t) + \int_0^1 u g_1(u) L(t, u) du = 1 . \tag{3.16}$$

It can be readily shown that for $h \to \infty$ the $L_1(t, u)$ and $L_2(t, u)$ tend to zero and the equation (3.16) reduces to the integral equation derived by MAL [4]. The terms corresponding to L_2 and L_3 account for the perturbations that will occur in MAL's [4] solution due to the finite boundary. If $\omega \ll 1$ it may be shown that

$$L_1(t, u) = o(\omega^2) ,$$

$$L_2(t, u) + L_3(t, u) = -\int_0^\infty [I_0(\xi t)\, I_0(\xi u)\, \{5 - 6\xi h + 2\xi^2 h^2 + 3 \exp(-2\xi h)\} + \\ + \{3 - 2\xi h + \exp(-2\xi h)\}\, \{u I_0(\xi t)\, I_1(\xi u) + \xi t I_0(\xi u)\, I_1(\xi t)\} + \\ + 2\xi^2 u t I_1(\xi u)\, I_1(\xi t)]\, [1 + o(\omega^2)] \frac{\xi\, d\xi}{2\xi h + \sinh(2\xi h)} .$$

If $\omega \to 0$, the dynamical problem considered here reduces to the statical problem studied by SNEDDON and SRIVASTAV [3] and the equation (3.16) reduces to the integral equation derived in [3].

The dynamical stress intensity factor defined by the equation

$$N = \lim_{x \to 1^+} (x - 1)^{1/2}\, |\sigma_{yy}(x, 0)| ,$$

is

$$N = \frac{p_0}{\sqrt{2}}\, g_1(l) \qquad (3.17)$$

The equation (3.16) has been solved numerically in the next section and the solution is used for calculating the dynamical stress intensity factor.

4. Numerical Solution of the Integral Equation

To compare our results with MAL [4] we vary k_2 in the range $0 \leq k_2 \leq 8$ and solve the equation (3.16) numerically by the method of FOX and GOODWIN [2]. The infinite integrals L_1, L_2 and L_3 were evaluated by using the five point Gauss-Laguerre quadrature formula. The only singularities of these integrals are branch points, associated with the values of α, α_1, β, and β_1 given in Section 2. A complex computer programme was used to evaluate the values of $g_1(t)$ and these values of $g_1(t)$ were used to calculate the values of the dynamic stress-intensity factor defined by (3.17). The results of these calculations are given in the table below and are illustrated graphically.

Table. Variation of $g_1(1)$ with h and k_2

k_2 / h	0	1	2	3	4	5	6	7	8
2	1.173	0.610	0.252	0.544	0.023	0.011	0.197	0.012	0.003
3	1.071	0.490	0.262	0.524	0.021	0.099	0.008	0.130	0.004
5	1.024	0.272	0.252	0.506	0.151	0.074	0.015	0.005	0.021

5. Discussion

When $k_2 = 0$, the problem reduces to the elastostatic problem studied by SNEDDON and SRIVASTAV [3]. The numerical values of the stress intensity factor obtained in this paper are in good agreement with those obtained by SNEDDON and SRIVASTAV for the static problem. MAL [4] studied the interaction of elastic waves with a Griffith crack situated in an infinite elastic medium. A comparison with MAL's results will yield valuable information about the effect of a finite boundary on the distribution of stresses. Inspection of [4], Fig. 2, shows that for the problem studied by MAL the stress intensity factor first increases with k_2 from its elastostatic value, reaches a maximum values, then decreases and exhibits an oscillatory behaviour. In the present case, see Fig., the stress intensity factor first decreases, reaches a minimum value, then increases. After reaching a maximum value at $k_2 = 3$, it then exhibits an oscillatory character similar to that found by MAL.

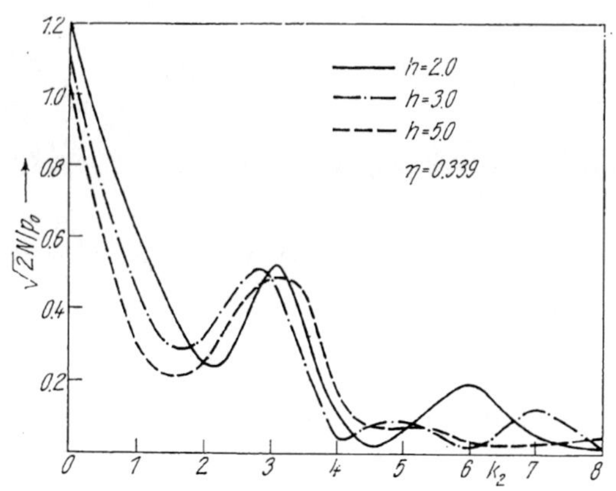

Fig.

The computations presented in this paper were performed on CDC 360-160-A computer system at Tata Institute of Fundamental Research Bombay. We are grateful to the Director of the institute for providing computer facilities. The author O. P. GUPTA is grateful to Professor G. M. L. GLADWELL, Department of Civil Engineering, University of Waterloo, Waterloo, Ontario, Canada for some suggestions for improving the paper.

Appendix

The following formulas were used in the analysis:

$$\text{(i)} \quad \frac{\mathrm{d}}{\mathrm{d}t} \int\limits_0^t (t^2 - x^2)^{-1/2}\, x \sinh(\alpha_1 x)\, \mathrm{d}x = \frac{\pi}{2} \alpha_1 t I_0^{\cdot}(\alpha_1 t)\,,$$

$$\text{(ii)} \quad \int\limits_0^\infty \xi(\xi^2 + \alpha_1^2)^{-1} \sin(\xi h)\, J_0(\xi u)\, \mathrm{d}\xi = \frac{\pi}{2} \alpha I_0(\alpha_1 u) \exp(-\alpha_1 h)\,,$$

$$\text{(iii)} \quad \int\limits_0^\infty (\xi^2 + \alpha_1^2)^{-1} \cos(\xi h)\, J_0(\xi u)\, \mathrm{d}\xi = \frac{\pi}{2} I_0(\alpha_1 u) \exp(-\alpha_1 h)\,.$$

These formulas can be found in [6].

References

1 SRIVASTAVA, K. N., GUPTA, O. P., and PALAIYA, R. M., Interaction of elastic waves in two bonded dissimilar elastic half-planes having Grffith crack at interface, Int. J. Fract. **14** (2) (1978).
2 FOX, L., and GOODWIN, E. T., The numerical solution of non-singular integral equations., Philos. Trans. Ser A., **245** (1953) 501.
3 SNEDDON, I. N., and SRIVASTAV, R. P., The stress in the vicinity of an infinite row of collinear cracks in an elastic body, Proc. Roy. Soc. Edin. A., **67** (1965) 39.
4 MAL, A. K., Interaction of elastic waves with a Griffith crack, Int. J. Engng. Sci., **8** (1970) 763.
5 SNEDDON, I. N., Mixed boundary value problems in potential theory, John Wiley (1966).
6 ERDELYI et al., Tables of Integral Transforms, Vol. 2, McGraw-Hill, New York (1954).

Eingereicht am 21. 11. 1980

Anschrift: Prof. K. N. SRIVASTAVA, Maulana Azad College of Technology, Bhopal-462007, Indien

ZAMM 61, 589 – 595 (1981)

G. S. GURAM / M. ANWAR

Micropolar Flow Due to a Rotating Disc with Suction and Injection

Es wird die inkompressible laminare stationäre Strömung einer mikropolaren Flüssigkeit betrachtet, welche von einer rotierenden Scheibe erzeugt wird. Dabei wird konstantes Einblasen oder Absaugen zugelassen. Die Bewegungsgleichungen werden dimensionslos gemacht, sie enthalten sechs Ähnlichkeitskennzahlen. Mit Hilfe einer Gauß-Seidel-Iteration werden die Gleichungen numerisch gelöst, es wird die Simpsonsche Regel verwendet. Einige Ergebnisse bei verschiedenen Werten des Absauge- bzw. Anblaseparameters werden in Tabellen und Diagrammen dargestellt und mit bekannten Ergebnissen für Newtonsche Flüssigkeiten verglichen.

The steady, laminar, and incompressible flow of a micropolar fluid due to a rotating disc with uniform suction and injection is considered. The equations of motion are reduced to dimensionless forms which include six dimensionless parameters, and integrated numerically by Gauss-Seidel iterative procedure and Simpson's rule. Some of the results are presented in the tables and diagrams, for different values of the suction and injection parameters, and compared with the known results for a Newtonian fluid.

Рассматривается стационарное несжимаемое и ламинарное течение микрополярной жидкости, вызванное вращением диска. Допускаются постоянное вдувание и отсосывание. Уравнения движения написаны в безразмерной форме. Они содержают шесть чисел подобия. Уравнения решены численным образом с помощью итерации Гаусса-Зейделя. При этом используется правило Симпсона. Некоторые результаты представлены в виде таблиц и диаграмм в зависимости от параметров вдувания и отсосывания. Результаты сравниваются с известными результатами для ньютоновских жидкостей.

Section 1. Introduction

ERINGEN [1] introduced a theory of micropolar fluids. These fluids are viscous fluids with five additional coefficients of viscosity when compared with the Newtonian fluids. They differ from non-Newtonian fluids in that they exhibit microinertial effects and can support couple stresses and body couples. The important feature of these fluids is the micro-rotation (or spin), which bears a resemblance to the vorticity in as much as only those components of it are non-vanishing which correspond to the non-vanishing components of vorticity and depend upon the same variable on which the vorticity components depend. Shearing stress components in these fluids are affected by the vorticity and micro-rotation of the fluid and are no longer symmetric. Physically, some polymetric fluids and fluids containing small amounts of polymeric additives may be represented by the mathematical model underlying micropolar fluids.

The problem of the steady flow due to a rotating disc for Newtonian fluids was first considered by KÁRMÁN [2]. COCHRAN [3] obtained a numerical solution for the same problem. The rotating disc problem with uniform suction and/or injection for Newtonian fluids has been considered by STUART [4], EVANS [5] and ACKROYD [6]. A numerical solution for the steady flow of a micropolar fluid due to a rotating disc without suction and injection, has been obtained by the authors [7].

In this paper, we consider the steady, laminar and incompressible flow of a micropolar fluid due to a rotating disc with uniform suction and injection. The basic equations of motion are stated in section 2, and the equations of motion are reduced to seven ordinary differential equations, in dimensionless form, involving six parameters. In section 3, the finite-difference equations are presented. The numerical procedure is described in section 4. In section 5, the results are presented both in tabular and graphical form, for different values of the suction and injection parameters and compared with the results obtained by STUART [4] and ACKROYD [6] for Newtonian fluids.

Section 2. Basic Equations and Analysis

The equations of motion of a micropolar fluid are [1]

$$\frac{\partial \varrho}{\partial t} + \nabla \cdot (\varrho \boldsymbol{v}) = 0\,, \tag{2.1}$$

$$(\lambda + 2\mu + \varkappa)\,\nabla(\nabla \cdot \boldsymbol{v}) - (\mu + \varkappa)\,\nabla \times \nabla \times \boldsymbol{v} + \varkappa\nabla \times \boldsymbol{\nu} - \nabla p + \varrho \boldsymbol{f} = \varrho \dot{\boldsymbol{v}}\,, \tag{2.2}$$

$$(\alpha + \beta + \gamma)\,\nabla(\nabla \cdot \boldsymbol{\nu}) - \gamma(\nabla \times \nabla \times \boldsymbol{\nu}) + \varkappa\nabla \times \boldsymbol{v} - 2\varkappa\boldsymbol{\nu} + \varrho \boldsymbol{l} = \varrho j \dot{\boldsymbol{\nu}}\,, \tag{2.3}$$

where ϱ is the density, $\boldsymbol{v}$ the velocity, $\boldsymbol{\nu}$ the micro-rotation or spin, p the thermo-dynamic pressure, $\boldsymbol{f}$ and $\boldsymbol{l}$ the body force and couple per unit mass, and j the micro-inertia; α, β, γ, λ, μ, and $\varkappa$ are material constants (viscosity coefficients); and the dot signifies material differentiation.

In the present work the steady, incompressible and axially-symmetric flow of a micropolar fluid, due to an infinite porous disc which is rotating with constant angular velocity Ω, is investigated. The cases of both uniform suction and injection are considered. We use cylindrical polar co-ordinates (r, θ, z) and denote the corresponding velocity and micro-rotation components by (u, v, w) and (ν_1, ν_2, ν_3) respectively. In the analysis, body forces and body couples are neglected. It is assumed that the material constants of the micropolar fluid are independent of position. We look for a solution for which

$$\boldsymbol{v} = \big(u(r,z), v(r,z), w(r,z)\big)\,, \qquad \boldsymbol{\nu} = \big(\nu_1(r,z), \nu_2(r,z), \nu_3(r,z)\big)\,. \tag{2.4}$$

For either suction or injection, it is assumed that on the disc $w = -\sqrt{k\Omega}A$, where $k = \dfrac{\mu + \varkappa}{\varrho}$. The parameter A is a dimensionless constant. We note that $A < 0$ corresponds to the injection case and $A > 0$ the suction case. The boundary conditions are

$$\begin{aligned} &z = 0\colon\ u = 0\,, \quad v = r\Omega\,, \quad w = -\sqrt{k\Omega}\,A\,, \quad \nu_1 = \nu_2 = \nu_3 = 0; \\ &z \to \infty\colon\ u = v = 0\,, \quad \nu_1 = \nu_2 = \nu_3 = 0\,, \quad p = 0\,. \end{aligned} \tag{2.5}$$

Using the dimensional analysis, the velocity, the micro-rotation, and the pressure are found to be of the form:

$$\begin{aligned} &u = r\Omega F_1(\eta)\,, \quad v = r\Omega F_2(\eta)\,, \quad w = \sqrt{k\Omega}\,F_3(\eta)\,, \\ &p = \varrho k\Omega P(\eta)\,, \quad \nu_1 = r\Omega\sqrt{\frac{\Omega}{k}}\,G_1(\eta)\,, \quad \nu_2 = r\Omega\sqrt{\frac{\Omega}{k}}\,G_2(\eta)\,, \quad \nu_3 = \Omega G_3(\eta)\,, \end{aligned} \tag{2.6}$$

where $\eta = z\sqrt{\dfrac{\Omega}{k}}$ being the dimensionless distance from the disc.

If we substitute (2.4) into the equations (2.1) through (2.3), we obtain the ordinary differential equations which after using the transformation $\xi = e^{-\eta}$, take the following form:

$$2F_1 - \xi F_3' = 0\,, \qquad \xi^2 F_1'' + \xi F_1' + \eta_1 \xi G_2' = -\xi F_3 F_1' - F_2^2 + F_1^2\,, \tag{2.7), (2.8}$$

$$\xi^2 F_2'' + \xi F_2' - \eta_1 \xi G_1' = -\xi F_3 F_2' + 2F_1 F_2\,, \qquad 2\xi F_1' + 2\eta_1 G_2 + \xi P' = -2F_1 F_3\,, \tag{2.9), (2.10}$$

$$\xi^2 G_1'' + \xi G_1' + \eta_2 \xi F_2' - 2\eta_2 G_1 = \eta_3 (F_1 G_1 - F_2 G_2 - \xi F_3 G_1')\,, \tag{2.11}$$

$$\xi^2 G_2'' + \xi G_2' - \eta_2 \xi F_1' - 2\eta_2 G_2 = \eta_3 (F_1 G_2 + F_2 G_1 - \xi F_3 G_2')\,, \tag{2.12}$$

$$\xi^2 G_3'' + \xi G_3' - \eta_4 \xi G_1' + \eta_5 (F_2 - G_3) = -\eta_6 \xi F_3 G_3'\,, \tag{2.13}$$

where the prime denotes differentiation with respect to ξ, and

$$\begin{aligned} &\eta_1 = \frac{\varkappa}{\mu + \varkappa}\,, \quad \eta_2 = \frac{\varkappa k}{\gamma \Omega}\,, \quad \eta_3 = \frac{k\varrho j}{\gamma}\,, \\ &\eta_4 = \frac{2(\alpha + \beta)}{\alpha + \beta + \gamma}\,, \quad \eta_5 = \frac{2\varkappa k}{\Omega(\alpha + \beta + \gamma)}\,, \quad \eta_6 = \frac{\varrho j k}{\alpha + \beta + \gamma}\,. \end{aligned} \tag{2.14}$$

The constants $\eta_1, \eta_2, \eta_3, \eta_4, \eta_5$ and η_6 are all dimensionless. The boundary conditions (2.5), in the independent variable ξ become as follows:

$$\begin{aligned} &\xi = 0;\ \ F_1 = F_2 = 0\,, \quad G_1 = G_2 = G_3 = 0\,, \quad P = 0; \\ &\xi = 1;\ F_1 = 0\,, \quad F_2 = 1\,, \quad F_3 = -A\,, \quad G_1 = G_2 = G_3 = 0\,. \end{aligned} \tag{2.15}$$

For a numerical approximation, the boundary conditions (2.15) are sufficient for any values of the parameters involved.

Section 3. Finite Difference Equations.

The interval $[0, 1]$ is divided into grid points with grid size h. A typical point is denoted by $\xi = \xi_i$. To approximate the equations (2.8), (2.9), (2.11), (2.12) and (2.13), we use the central-differences for all the derivatives involved. Thus the following finite-difference equations are obtained.

$$\begin{aligned} &(2\xi_i^2 - h\xi_i)\,F_{i-1}^{(1)} - 4\xi_i^2 F_i^{(1)} + (2\xi_i^2 + h\xi_i)\,F_{i+1}^{(1)} = \\ &= -\eta_1 \xi_i h(G_{i+1}^{(2)} - G_{i-1}^{(2)}) - h\xi_i F_i^{(3)}(F_{i+1}^{(1)} - F_{i-1}^{(1)}) - 2h^2 F_i^{(2)} F_i^{(2)} + 2h^2 F_i^{(1)} F_i^{(1)}\,, \end{aligned} \tag{3.1}$$

$$\begin{aligned} &(2\xi_i^2 - h\xi_i)\,F_{i-1}^{(2)} - 4\xi_i^2 F_i^{(2)} + (2\xi_i^2 + h\xi_i)\,F_{i+1}^{(2)} = \\ &= \eta_1 \xi_i h(G_{i+1}^{(1)} - G_{i-1}^{(1)}) - \xi_i h F_i^{(3)}(F_{i+1}^{(2)} - F_{i-1}^{(2)}) + 4h^2 F_i^{(1)} F_i^{(2)}\,, \end{aligned} \tag{3.2}$$

$$\begin{aligned} &(2\xi_i^2 - h\xi_i)\,G_{i-1}^{(1)} - (4\xi_i^2 + 4\eta_2 h^2)\,G_i^{(1)} + (2\xi_i^2 + h\xi_i)\,G_{i+1}^{(1)} = \\ &= -\eta_2 \xi_i h(F_{i+1}^{(2)} - F_{i-1}^{(2)}) + \eta_3[2h^2 F_i^{(1)} G_i^{(1)} - 2h^2 F_i^{(2)} G_i^{(2)} - \xi_i h F_i^{(3)}(G_{i+1}^{(1)} - G_{i-1}^{(1)})\,, \end{aligned} \tag{3.3}$$

$$\begin{aligned} &(2\xi_i^2 - h\xi_i)\,G_{i-1}^{(2)} - 4(\xi_i^2 + \eta_2 h^2)\,G_i^{(2)} + (2\xi_i^2 + h\xi_i)\,G_{i+1}^{(2)} = \\ &= \eta_2 h\xi_i(F_{i+1}^{(1)} - F_{i-1}^{(1)}) + \eta_3[2h^2 F_i^{(1)} G_i^{(2)} + 2h^2 F_i^{(2)} G_i^{(1)} - h\xi_i F_i^{(3)}(G_{i+1}^{(2)} - G_{i-1}^{(2)})]\,, \end{aligned} \tag{3.4}$$

$$\begin{aligned} &(2\xi_i^2 - h\xi_i)\,G_{i-1}^{(3)} - 2(2\xi_i^2 + \eta_5 h^2)\,G_i^{(3)} + (2\xi_i^2 + h\xi_i)\,G_{i+1}^{(3)} = \\ &= \eta_4 h\xi_i(G_{i+1}^{(1)} - G_{i-1}^{(1)}) - 2\eta_5 h^2 F_i^{(2)} - \eta_6 \xi_i h F_i^{(3)}(G_{i+1}^{(3)} - G_{i-1}^{(3)})\,. \end{aligned} \tag{3.5}$$

which are solved for all interior points of the interval [0, 1] with the conditions (2.15) for F_1, F_2, G_1, G_2 and G_3 at the boundary points and the first-order differential equations (2.7) and (2.10) are integrated directly, with the conditions on F_3 and P, by the method discussed in the next section.

Section 4. Numerical Procedure

To solve the above system of finite-difference equations, subject to the boundary conditions on F_1, F_2, G_1, G_2 and G_3, the Gauss-Seidel iterative procedure [8, p. 154—156] is used. The equations (2.7) and (2.10) are solved using the Simpson's rule [8, p. 67—69] with the formula given in [9, p. 48] and the boundary conditions on F_3 and P.

The sequence of iteration is given below:

(i) The equations (3.1) are solved with $F_1 = 0$ when $\xi = 0$ and $\xi = 1$. To calculate the elements of the matrix associated with F_1, the most recently available values of F_1, F_2, F_3 and G_2 were used. The first-order differential equation (2.7) is then integrated by Simpson's rule with the formula given in [9, p. 48] subject to the boundary conditions on F_3, using the computed value of F_1 in the calculations for F_3;

(ii) the calculated values for F_1 and F_3 are introduced into (3.2) and these equations are solved subject to $F_2 = 0$ at $\xi = 0$, $F_2 = 1$ at $\xi = 1$. The most recently available information for F_1, F_2, F_3 and G_1 is again used for calculating the values of F_2;

(iii) the computed solutions for F_1, F_2 and F_3 are employed into (3.3), (3.4) and (3.5) and these sets of equations are solved for G_1, G_2 and G_3 respectively, subject to $G_1 = G_2 = G_3 = 0$ when $\xi = 0$ and $\xi = 1$, using the recent values for F_1, F_2, F_3, G_1, G_2 and G_3. The solutions for these equations give updated information for the subsequent application of (i) and (ii).

The above precedure is repeated until all the solutions have approached to some required criterion of accuracy. The criterion

$$|F_i^{(n+1)} - F_i^{(n)}| < 10^{-6}, \qquad |G_i^{(n+1)} - G_i^{(n)}| < 10^{-6}, \qquad i = 1, 2, 3$$

was adopted as a terminating condition, where here the subscript denotes the number of a function and the superscript the iteration.

The computed solutions of F_1, F_3 and G_2 were introduced to solve (2.10) to calculate the pressure with $P = 0$ when $\xi = 0$.

It must be noticed that to calculate the values of F_3 and the right hand side of the equation involving pressure namely,

$$P' + 2F_1' = -\frac{2}{\xi}(F_1F_3 + \eta_1 G_2)$$

at $\xi = 0$ only, the first derivative was approximated by forward differences after the application of l'Hôpital's rule.

Section 5. Results and Discussion

The results are computed for the following three sets of the parameters for different values of injection and suction:

	η_1	η_2	η_3	η_4	η_5	η_6
(a)	0.4	0.9	0.2	0.1	1.0	0.4
(b)	0.8	1.8	0.4	0.2	2.0	0.8
(c)	1.2	2.7	0.6	0.3	3.0	1.2

The four cases of injection and suction considered are as follows: Case 1. $A = -0.5$; Case 2. $A = -0.2$; Case 3. $A = 1.0$; Case 4. $A = 3.0$. The first two cases correspond to the injection, whereas the last two the suction.

For each above case, the results have been obtained for all the three sets (a), (b), and (c) of the parameters. But the detailed results are presented for F_1, F_2, F_3, G_1, G_2, G_3 and P in tables 1—4 only for the cases 1(a), 2(b), 3(c) and 4(a) respectively, in order to curb the length of the paper. The numerical solutions have been carefully checked by performing calculations on different grid sizes namely (i) $h = 0.025$ (ii) $h = 0.0125$ from which an estimate of the accuracy may be obtained. The trend of these results as the grid size is decreased tends to indicate that the solutions for $h = 0.0125$ are accurate to within a 0.06% tolerance for F_1, F_2, G_1, G_2, G_3, and 0.3% for F_3 and P approximately.

Typical sets of solutions, showing the effect of the suction and injection parameter A are given in graphical form in figures 1—7 for the set (a) of the parameters η_1, η_2, η_3, η_4, η_5 and η_6. The case without suction or injection as given in [7] is also shown in the graphs for radial (F_1) and tangential (F_2) profiles.

It is noticed that the effect of suction shows a rapid decrease in magnitude of the radial component with the increase of suction and the axial flow at infinity toward the disc is larger. The profiles of F_2, G_1, G_2, G_3 and P are lowered as well, as the suction increases. In the case of injection, all the profiles of F_1, F_2, F_3, G_1, G_2, G_3 and P are raised with the increase in injection. The comparison of the results is made with the results of the Newtonian fluid given by ACKROYD [6] in table 5 and graphically for F_1 and F_2 with STUART [4] in figures 8 and 9, and in table 6 for F_1 and F_3. Figures 8 and 9 indicate that for micropolar fluids F_1 and F_2 are less than that of the Newtonian fluids.

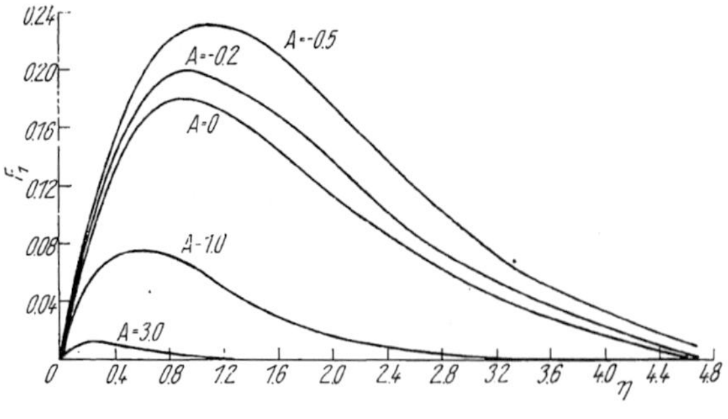

Fig. 1

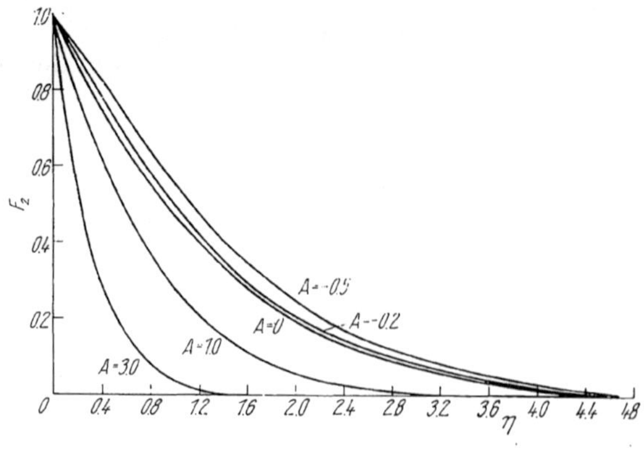

Fig. 2

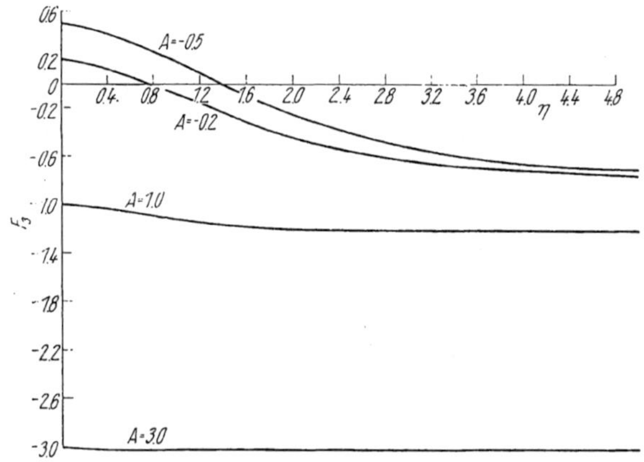

Fig. 3

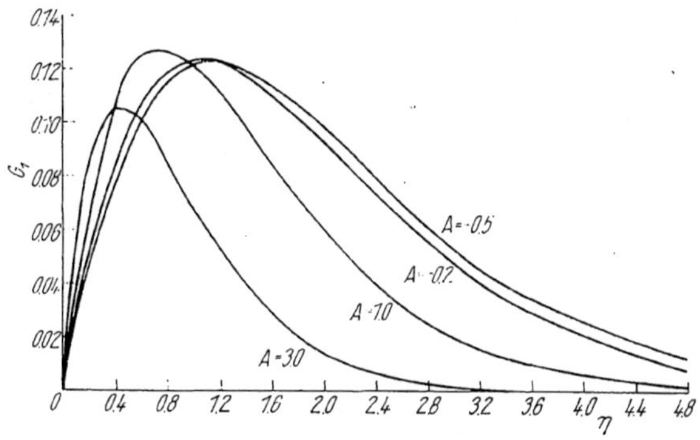

Fig. 4

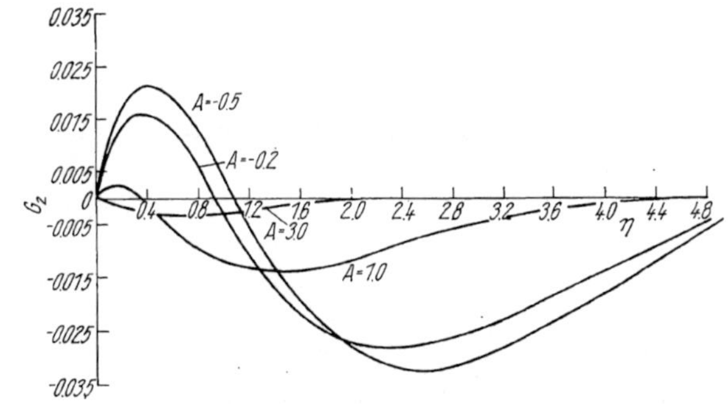

Fig. 5

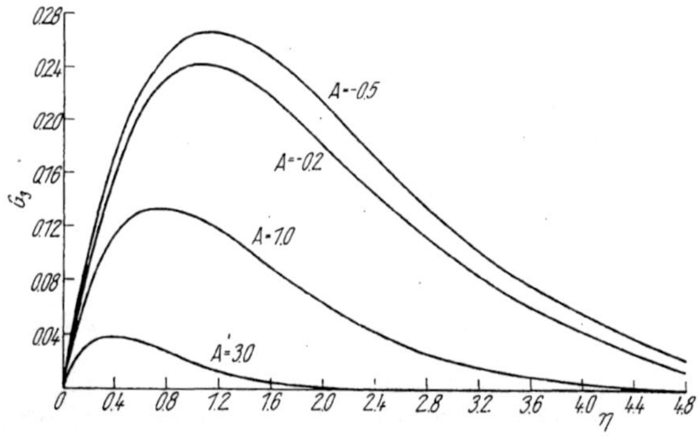

Fig. 6

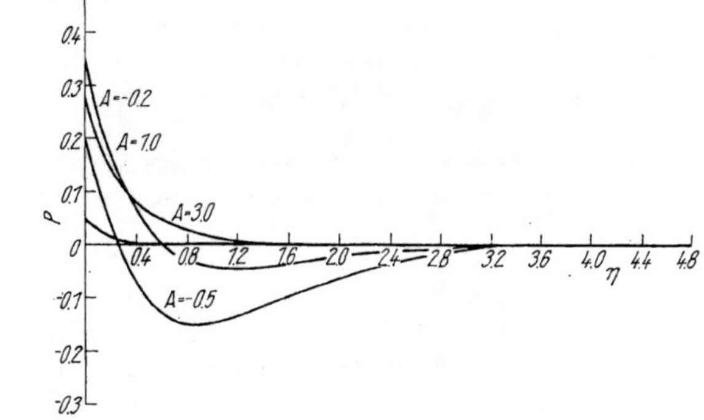

Fig. 7

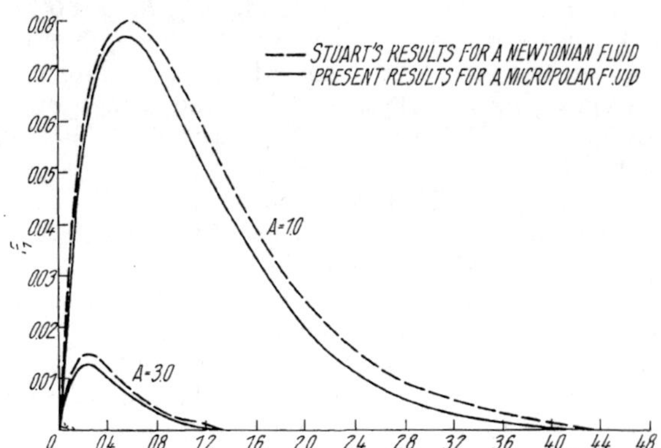

Fig. 8

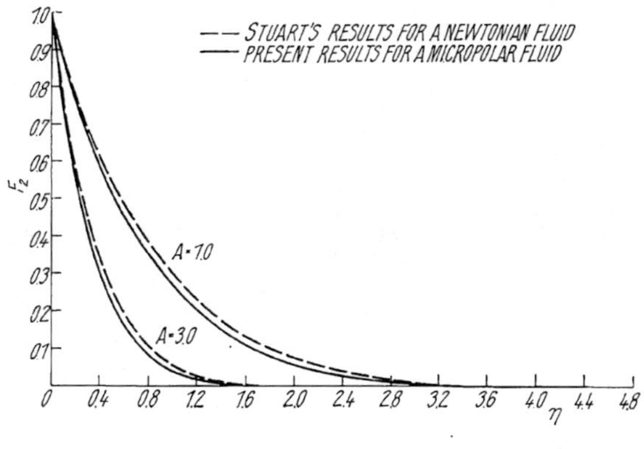

Fig. 9

Table 1. Case 1 (a)

grid size	ξ	F_1	F_2	F_3	G_1	G_2	G_3	P
$\frac{1}{40}$	0.0	0.0000000	0.0000000	−0.7131647	0.0000000	0.0000000	0.0000000	0.0000000
	0.1	0.1500217	0.1930444	−0.3435520	0.0836078	−0.0318601	0.1827571	−0.0521829
	0.2	0.2119364	0.3482124	−0.0913995	0.1133003	−0.0203251	0.2475852	−0.1058902
	0.3	0.2330274	0.4789096	0.0904958	0.1222634	−0.0053180	0.2676393	−0.1436843
	0.4	0.2296389	0.5906947	0.2245735	0.1191957	0.0074352	0.2618341	−0.1582668
	0.5	0.2101956	0.6869097	0.3233685	0.1083412	0.0163544	0.2391693	−0.1486026
	0.6	0.1797374	0.7698960	0.3949081	0.0921833	0.0210478	0.2047624	−0.1161477
	0.7	0.1415404	0.8414231	0.4447508	0.0723218	0.0215456	0.1618153	−0.0633468
	0.8	0.0978532	0.9028897	0.4769494	0.0498515	0.0180377	0.1124572	0.0070359
	0.9	0.0502802	0.9554335	0.4945705	0.0255530	0.0107699	0.0581658	0.0922298
	1.0	0.0000000	1.0000000	0.5000000	0.0000000	0.0000000	0.0000000	0.1896046
$\frac{1}{80}$	0.0	0.0000000	0.0000000	−0.7246543	0.0000000	0.0000000	0.0000000	0.0000000
	0.1	0.1505993	0.1928408	−0.3443701	0.0835479	−0.0315485	0.1828075	−0.0430268
	0.2	0.2122024	0.3480200	−0.0916406	0.1132742	−0.0200614	0.2475556	−0.0960187
	0.3	0.2331567	0.4787571	0.0904115	0.1222562	−0.0051146	0.2675988	−0.1335863
	0.4	0.2297021	0.5905787	0.2245432	0.1191975	0.0075898	0.2617972	−0.1480789
	0.5	0.2102252	0.6868235	0.3233583	0.1083467	0.0164704	0.2391394	−0.1383753
	0.6	0.1797499	0.7698338	0.3949053	0.0921899	0.0211325	0.2047399	−0.1059030
	0.7	0.1415444	0.8413807	0.4447504	0.0723279	0.0216043	0.1617997	−0.0530950
	0.8	0.0978536	0.9028637	0.4769495	0.0498561	0.0180743	0.1124475	0.0172895
	0.9	0.0502797	0.9554214	0.4945705	0.0255555	0.0107871	0.0581614	0.1024828
	1.0	0.0000000	1.0000000	0.5000000	0.0000000	0.0000000	0.0000000	0.1998559

Table 2. Case 2 (b)

grid size	ξ	F_1	F_2	F_3	G_1	G_2	G_3	P
$\frac{1}{40}$	0.0	0.0000000	0.0000000	−0.7142123	0.0000000	0.0000000	0.0000000	0.0000000
	0.1	0.0989016	0.1281182	−0.4994362	0.0773547	−0.0350107	0.1358236	0.0040484
	0.2	0.1633782	0.2729240	−0.3185242	0.1289098	−0.0320493	0.2363677	−0.0116403
	0.3	0.1953264	0.4080415	−0.1720689	0.1557712	−0.0167927	0.2937381	−0.0231229
	0.4	0.2030531	0.5299988	−0.0565761	0.1640234	0.0006309	0.3156690	−0.0214898
	0.5	0.1928314	0.6384117	0.0324196	0.1579683	0.0155310	0.3091075	−0.0028761
	0.6	0.1692852	0.7337478	0.0989185	0.1407369	0.0255370	0.2794831	0.0337665
	0.7	0.1358743	0.8167529	0.1463076	0.1146865	0.0294277	0.2309440	0.0878783
	0.8	0.0952349	0.8882522	0.1774248	0.0816383	0.0266011	0.1666569	0.1580058
	0.9	0.0494087	0.9490707	0.1946521	0.0430271	0.0168068	0.0890508	0.2422348
	1.0	0.0000000	1.0000000	0.2000000	0.0000000	0.0000000	0.0000000	0.3384572
$\frac{1}{80}$	0.0	0.0000000	0.0000000	−0.6957883	0.0000000	0.0000000	0.0000000	0.0000000
	0.1	0.0982521	0.1272492	−0.4980907	0.0778149	−0.0351602	0.1348415	0.0006979
	0.2	0.1629611	0.2725847	−0.3179148	0.1292349	−0.0321511	0.2358275	−0.0157936
	0.3	0.1950346	0.4079104	−0.1717462	0.1559851	−0.0168726	0.2934066	−0.0276066
	0.4	0.2028434	0.5299563	−0.0563973	0.1641617	0.0005649	0.3154525	−0.0261497
	0.5	0.1926806	0.6384069	0.0325181	0.1580561	0.0154762	0.3089618	−0.0076456
	0.6	0.1691791	0.7337571	0.0989703	0.1407908	0.0254926	0.2793848	0.0289202
	0.7	0.1358032	0.8167650	0.1463322	0.1147177	0.0293937	0.2308798	0.0829733
	0.8	0.0951920	0.8882619	0.1774342	0.0816544	0.0265779	0.1666187	0.1530523
	0.9	0.0493892	0.9490759	0.1946541	0.0430333	0.0167949	0.0890334	0.2372392
	1.0	0.0000000	1.0000000	0.2000000	0.0000000	0.0000000	0.0000000	0.3334241

Table 3. Case 3 (c)

grid size	ξ	F_1	F_2	$-F_3$	G_1	G_2	G_3	P
1/40	0.0	0.0000000	0.0000000	1.1208710	0.0000000	0.0000000	0.0000000	0.0000000
	0.1	0.0005009	−0.0077330	1.1236418	0.0168492	−0.0037424	0.0016343	−0.0009665
	0.2	0.0098012	0.0315535	1.1181020	0.0741910	−0.0120297	0.0307363	−0.0010548
	0.3	0.0243387	0.1132458	1.1048579	0.1378707	−0.0181919	0.0800770	−0.0005994
	0.4	0.0397524	0.2232203	1.0865797	0.1904610	−0.0199136	0.1303664	0.0020010
	0.5	0.0523385	0.3502088	1.0659707	0.2232895	−0.0171475	0.1703089	0.0091412
	0.6	0.0592128	0.4856533	1.0454617	0.2319318	−0.0111608	0.1918842	0.0233616
	0.7	0.0583315	0.6229850	1.0271200	0.2142735	−0.0039791	0.1893200	0.0469675
	0.8	0.0484394	0.7570708	1.0126242	0.1695660	0.0019791	0.1584180	0.0818109
	0.9	0.0289817	0.8838342	1.0032711	0.0979185	0.0041516	0.0960864	0.1291739
	1.0	0.0000000	1.0000000	1.0000000	0.0000000	0.0000000	0.0000000	0.1897205
1/80	0.0	0.0000000	0.0000000	1.1214146	0.0000000	0.0000000	0.0000000	0.0000000
	0.1	0.0005203	−0.0068816	1.1237023	0.0166723	−0.0037261	−0.0007686	−0.0006370
	0.2	0.0097974	0.0320880	1.1181578	0.0740516	−0.0120511	0.0313375	−0.0006701
	0.3	0.0243465	0.1135049	1.1049132	0.1377695	−0.0182296	0.0805111	−0.0002068
	0.4	0.0397746	0.2234719	1.0866264	0.1903858	−0.0199518	0.1306888	0.0024019
	0.5	0.0523698	0.3503881	1.0660053	0.2232322	−0.0171780	0.1705504	0.0095566
	0.6	0.0592465	0.4857800	1.0454843	0.2318878	−0.0111806	0.1920627	0.0237967
	0.7	0.0583617	0.6230712	1.0271326	0.2142403	−0.0039888	0.1894464	0.0474256
	0.8	0.0484619	0.7571238	1.0126296	0.1695431	0.0019767	0.1584989	0.0822938
	0.9	0.0289937	0.8838591	1.0032724	0.0979063	0.0041527	0.0961258	0.1296822
	1.0	0.0000000	1.0000000	1.0000000	0.0000000	0.0000000	0.0000000	0.1902538

Table 4. Case 4 (a)

grid size	ξ	F_1	F_2	$-F_3$	G_1	G_2	G_3	P
1/40	0.0	0.0000000	0.0000000	3.0161721	0.0000000	0.0000000	0.0000000	0.0000000
	0.1	−0.0000413	−0.0014634	3.0162614	0.0085983	−0.0003204	0.0009046	−0.0000308
	0.2	0.0001587	0.0015020	3.0162450	0.0276675	−0.0010005	0.0060778	−0.0000467
	0.3	0.0009587	0.0169448	3.0158642	0.0512499	−0.0017770	0.0141966	−0.0000559
	0.4	0.0026191	0.0517744	3.0149030	0.0746906	−0.0024330	0.0234320	0.0000068
	0.5	0.0052166	0.1123258	3.0132089	0.0938711	−0.0027852	0.0319175	0.0003880
	0.6	0.0084949	0.2045529	3.0107396	0.1049972	−0.0027068	0.0378111	0.0016754
	0.7	0.0116737	0.3341232	3.0076251	0.1045010	−0.0021664	0.0393092	0.0049956
	0.8	0.0132227	0.5064639	3.0042456	0.0889867	−0.0012779	0.0346490	0.0122434
	0.9	0.0106059	0.7267736	3.0013190	0.0551980	−0.0003593	0.0221075	0.0263411
	1.0	0.0000000	1.0000000	3.0000000	0.0000000	0.0000000	0.0000000	0.0515197
1/80	0.0	0.0000000	0.0000000	3.0162251	0.0000000	0.0000000	0.0000000	0.0000000
	0.1	−0.0000382	−0.0013890	3.0163025	0.0085924	−0.0003210	0.0009582	−0.0000089
	0.2	0.0001644	0.0016255	3.0162802	0.0276541	−0.0010025	0.0061486	−0.0000117
	0.3	0.0009674	0.0171058	3.0158938	0.0512270	−0.0017804	0.0142744	−0.0000088
	0.4	0.0026311	0.0519611	3.0149266	0.0746582	−0.0024377	0.0235096	0.0000662
	0.5	0.0052314	0.1125250	3.0132266	0.0938309	−0.0027907	0.0319892	0.0004608
	0.6	0.0085112	0.2047498	3.0107515	0.1049527	−0.0027126	0.0378728	0.0017633
	0.7	0.0116894	0.3343011	3.0076321	0.1044570	−0.0021719	0.0393577	0.0051004
	0.8	0.0132352	0.5066040	3.0042488	0.0889495	−0.0012824	0.0346822	0.0123665
	0.9	0.0106130	0.7268552	3,0013198	0.0551750	−0.0003621	0.0221243	0.0264827
	1.0	0.0000000	1.0000000	3.0000000	0.0000000	0.0000000	0.0000000	0.0516777

Table 5. Comparison of present results with ACKROYD's [6] results

	A	$F_1'(0)$	$-F_2'(0)$	$P(0)$
Present results for a micropolar fluid	−0.5	0.5120344	0.4131584	0.1998559
	−0.2	0.5128488	0.5136968	0.3383950
	1.0	0.3741984	1.1579912	0.2722234
	3.0	0.1545296	2.9758048	0.0516777
ACKROYD's results for a Newtonian fluid	−0.5	0.5145663	0.4364319	0.2007953
	−0.2	0.5162650	0.5379560	0.3399210
	1.0	0.3895662	1.1752208	0.2944971
	3.0	0.1655778	3.0121419	0.0547905

Table 6. Comparison of present results with STUART's [4] results

	A	F_1 (maximum)	$-F_3(\infty)$
Present results for a micropolar fluid	1.0	0.077	1.224
	3.0	0.0132	3.016
STUART's results for a Newtonian fluid	1.0	0.080	1.259
	3.0	0.0136	3.019

Acknowledgement

The financial support from NSERC (grant ∓A3497) is gratefully acknowledged.

References

1 ERINGEN, A. C., Theory of micropolar fluids, J. Math. Mech., **16**, (1966), 1—18.
2 VON KÁRMÁN, TH., Über laminare und turbulente Reibung, Zeitschr. f. Angew. Math. u. Mech., **1** (1921), 233—252.
3 COCHRAN, W. G., The flow due to a rotating disc, Proc. Camb. Phil. Soc. **30**, (1934), 365—375.
4 STUART, J. T., On the effects of uniform suction on the steady flow due to a rotating disk, Quart, J. Mech. Appl. Math., **7**, (1954), 446—457.
5 EVANS, D. J., The rotationally symmetric flow of a viscous fluid in the presence of an infinite rotating disc with uniform suction, Quart, J. Mech. Appl. Math., **22**, (1969), 467—485.
6 ACKROYD, J. A. D., On the steady flow produced by a rotating disc with either surface suction or injection, J. Engng. Math., **12**, (1978), 207—220.
7 GURAM, G. S.; ANWAR, M., Steady flow of a micropolar fluid due to a rotating disc, J. Engng. Math., **13**, (1979), 223—234.
8 GERALD, C. F., Applied Numerical Analysis, Addison-Wesly Publishing Company, Reading, Massachusetts (1974).
9 MILNE, W. E., Numerical solution of differential equations, John Wiley & Sons Inc., New York (1953).

Eingereicht am 27. 8. 1980, in revidierter Fassung am 30. 12. 1980

Anschrift: G. S. GURAM, Assistant Professor, University of Western Ontario, Faculty of Science, Dept. of Appl. Math., London, Canada, N6A 5B9

KLEINE MITTEILUNGEN

ZAMM 61, 597 – 598 (1981)

J. PERKKIÖ / R. KESKINEN

On the Velocity Profiles of Newtonian Suspension in Cylindrical Tube

In this paper we have studied the steady flow of a Newtonian suspension in a cylindrical tube and calculated explicit expressions for the velocity profiles in some cases of changing viscosity. This problem arose from studies concerning the flow of blood in a bifurcating tube.

The constitutive equation of a Newtonian fluid is

$$\Sigma = -pI + 2\mu D\,, \tag{1}$$

where Σ is the stress tensor, p the isotropic pressure, μ the dynamic viscosity and D the rate of deformation tensor. We have supposed that the viscosity depends on the concentration of the suspended particles but not on the flow conditions, i.e. $\mu = \mu(x)$.

Cauchy's first law of motion is

$$\varrho\,\frac{\mathrm{d}\bar v}{\mathrm{d}t} = \bar F_B + \nabla\cdot\Sigma^T\,, \tag{2}$$

where $\bar F_B$ is the density of body forces and Σ^T is the transpose of the stress tensor. Substituting Eq. (1) into Eq. (2) and using the expression

$$D = \frac{1}{2}\,(\nabla\bar v + (\nabla\bar v)^T) \tag{3}$$

the following equation is obtained:

$$\varrho\,\frac{\mathrm{d}\bar v}{\mathrm{d}t} = \bar F_B - \nabla p + \nabla\mu\cdot\nabla\bar v + \nabla\bar v\cdot\nabla\mu + \mu(\nabla^2\bar v + \nabla(\nabla\cdot\bar v))\,. \tag{4}$$

In the cylindrical coordinates $\mu = \mu(r)$, $p = p(z)$ and $\bar v = v(r)\,\bar e_z$ as the z-axis is along the tube axis. Supposing that the suspension is incompressible ($\nabla\cdot\bar v = 0$) and that the flow is steady $\left(\frac{\mathrm{d}\bar v}{\mathrm{d}t} = 0\right)$ in a horizontal tube ($\bar F_B = \varrho\bar g$ is perpendicular to the z-axis) the equation

$$-\frac{\mathrm{d}p}{\mathrm{d}z} + \frac{\mathrm{d}\mu}{\mathrm{d}r}\,\frac{\mathrm{d}v}{\mathrm{d}r} + \mu\,\frac{1}{r}\,\frac{\mathrm{d}}{\mathrm{d}r}\left(r\,\frac{\mathrm{d}v}{\mathrm{d}r}\right) = 0 \tag{5}$$

is obtained.

Defining $\xi = r/R$ and $f(\xi) = \xi\,\frac{\mathrm{d}v}{\mathrm{d}\xi}$ from Eq. (5) follows

$$\frac{\mathrm{d}}{\mathrm{d}\xi}\,(\mu f) = -\frac{\Delta pR^2}{L}\,\xi\,, \tag{6}$$

where $\Delta p/L$ is the absolute value of the pressure gradient. The boundary conditions are $f(0) = 0$ and $v(1) = 0$, so the integration yields

$$v(\xi) = \frac{\Delta pR^2}{2L}\int\limits_{\xi}^{1}\mathrm{d}\xi\,\frac{\xi}{\mu(\xi)}\,. \tag{7}$$

Different expressions for the velocity profiles are obtained from Eq. (7).

1. The profile is parabolic in the case of a constant viscosity

$$v(\xi) = \frac{\Delta pR^2}{4\mu L}\,(1 - \xi^2)\,, \tag{8}$$

which is the Poiseuille equation.

2. The modified Einstein equation for dilute suspensions is

$$\mu(\xi) = \mu_0(1 + Kc_V(\xi)) \tag{9}$$

where $\mu(\xi)$ and μ_0 are the viscosities of the whole suspension and the suspending fluid, respectively, K is the geometric factor ($K \geqq 2.5$) and $c_V(\xi)$ is the volume concentration of the suspended particles.

We have supposed that the concentration depends linearly on the radial coordinate, i.e.

$$c_V(\xi) = c_V(1 - \xi)\,. \tag{10}$$

The velocity profile obtained from Eq. (7) is

$$v(\xi) = \frac{\Delta pR^2}{2\mu_0 L}\,\frac{1}{Kc_V}\left\{\xi - 1 + \frac{1 + Kc_V}{Kc_V}\ln\left(1 + Kc_V(1 - \xi)\right)\right\}. \tag{11}$$

3. In the case of blood a reasonable approximation for the viscosity-concentration dependence is (WHITMORE [1])

$$\mu = \mu_0\,\mathrm{e}^{B'c_V(\xi)} = \mu_0\,\mathrm{e}^{B(1-\xi)}\,, \tag{12}$$

where $B \approx 1.2$. The integration yields from Eq. (7)

$$v(\xi) = \frac{\Delta pR^2}{2\mu_0 L}\,\mathrm{e}^{-B}\left\{\frac{1}{B^2}\,(\mathrm{e}^{B\xi} - \mathrm{e}^{B}) - \frac{1}{B}\,(\xi\,\mathrm{e}^{B\xi} - \mathrm{e}^{B})\right\}. \tag{13}$$

In the limit $c_V(\xi) \to 0$ ($B \to 0$) the solutions (11) and (13) must approach the Poiseuille equation, because $\mu(\xi) \to \mu_0$. Using the series expansions of the logarithm and the exponential function the parabolic profile is easily obtained as $c_V(\xi) \to 0$.

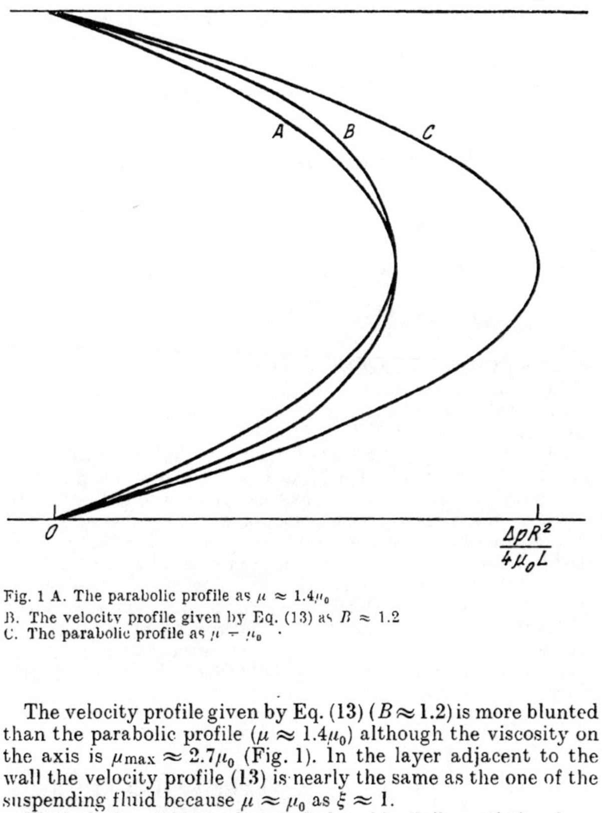

Fig. 1 A. The parabolic profile as $\mu \approx 1.4\mu_0$
B. The velocity profile given by Eq. (13) as $B \approx 1.2$
C. The parabolic profile as $\mu = \mu_0$

The velocity profile given by Eq. (13) ($B \approx 1.2$) is more blunted than the parabolic profile ($\mu \approx 1.4\mu_0$) although the viscosity on the axis is $\mu_{\max} \approx 2.7\mu_0$ (Fig. 1). In the layer adjacent to the wall the velocity profile (13) is nearly the same as the one of the suspending fluid because $\mu \approx \mu_0$ as $\xi \approx 1$.

The solution (13) can be applied to blood flows if the shear rates are high enough that blood will behave like a Newtonian fluid.

By changing the boundary conditions the result (13) can easily be modified to the flows in which the concentration is zero in the particle-poor layer adjacent to the wall and changes linearly in a thin layer towards a constant value. Such a situation is more reasonable than the linear change of concentration across the whole cross-section of the tube. However, the chosen concentration profile (10) reveals the general features of the solutions given by Eq. (7) more clearly than the piecewise linear concentration change.

References

1 WHITMORE, R. L., Rheology of the circulation, Pergamon Press, Oxford, (1968), p. 79.

Eingereicht am 20. 3. 1981

Anschriften: M. Sc. J. PERKKIÖ, Department of Medical Physics, University of Helsinki, Siltavuorenpenger 10, 00170 Helsinki 17, Finland. – Dr. R. KESKINEN, Department of Theoretical Physics, University of Helsinki, Siltavuorenpenger 20, 00170 Helsinki 17, Finland.

ZAMM 61, 598 – 599 (1981)

M. E. MULDOON

A Differential Equations Proof of a Nicholson-Type Formula

Introduction

G. N. WATSON [11, ch. 13] proves Nicholson's formula

$$J_\nu^2(z) + Y_\nu^2(z) = 8\pi^{-2} \int_0^\infty K_0(2z \sinh t) \cosh 2\nu t \, dt\,, \qquad \operatorname{Re} z > 0\,, \tag{1}$$

for the sum of squares of Bessel functions and a companion formula

$$J_\nu(z)\, \partial Y_\nu(z)/\partial\nu - Y_\nu(z)\, \partial J_\nu(z)/\partial\nu = -4\pi^{-1} \int_0^\infty K_0(2z \sinh t)\, e^{-2\nu t}\, dt\,, \qquad \operatorname{Re} z > 0\,, \tag{2}$$

by the methods of complex integration. Formula (2) is useful in showing that

$$J_\nu(\nu)\, d Y_\nu(\nu)/d\nu - Y_\nu(\nu)\, dJ_\nu(\nu)/d\nu > 0$$

[11, p. 448] and in other places; see, e.g., [8]. As an almost immediate consequence of (2), WATSON derives a formula

$$dc/d\nu = 2c \int_0^\infty K_0(2c \sinh t)\, e^{-2\nu t}\, dt \tag{3}$$

for the derivative with respect to order of a zero of a fixed linear combination $C_\nu(z)$ of $J_\nu(z)$ and $Y_\nu(z)$. This last formula has been found useful in many investigations of zeros of Bessel and related functions; see, e.g., [2, 3, 4, 5, 6, 7, 10, 13].

J. E. WILKINS, Jr. [12] has given a simpler "differential equations" proof of (1). His proof is reproduced in a modified form in [9, pp. 340–341]. The purpose of the present note is to give a "differential equations" proof of (2). The principal part of WILKINS' method consisted in showing that both sides of (1) satisfy the same third-order homogeneous linear differential equation. In the present situation we show that both sides of (2) satisfy the same third-order nonhomogeneous linear differential equation. The work is shortened by making the change of variable $\theta = \log z$ where the principal value of the logarithm is intended. WILKINS' proof could be shortened a little by this observation also.

Proof of (2)

We consider first the case $\nu \neq 0$. We may write (2) as

$$J_\nu(e^\theta)\, \partial Y_\nu(e^\theta)/\partial\nu - Y_\nu(e^\theta)\, \partial J_\nu(e^\theta)/\partial\nu = -\frac{4}{\pi} \int_0^\infty K_0(2\, e^\theta \sinh t)\, e^{-2\nu t}\, dt\,, \qquad \theta \in S\,, \tag{4}$$

where S is the strip $|\operatorname{Im} \theta| < \pi/2$ in the θ-plane.

We see from the asymptotic behaviour of $K_0(2z \sinh t)$ as $t \to 0^+$ and ∞, respectively, that the integral in (4) converges uniformly in any compact subset of S. Hence [1, p. 110] it defines a function which may be differentiated indefinitely in S. We write $z(\theta)$ for the right-hand side of (4).

We remark that the differential equation for K_0 shows that

$$\frac{d}{dt} [2\, e^\theta \sinh t K_0'(2\, e^\theta \sinh t)] = 2\, e^{2\theta} \sinh 2t K_0(2 e^\theta \sinh t) \tag{5}$$

and

$$\frac{d}{d\theta} [2\, e^\theta \sinh t K_0'(2\, e^\theta \sinh t)] = 4\, e^{2\theta} \sinh^2 t K_0(2\, e^\theta \sinh t)\,. \tag{6}$$

We have

$$z'(\theta) = -\frac{4}{\pi} \int_0^\infty K_0'(2\, e^\theta \sinh t)\, 2 e^\theta \sinh t\, e^{-2\nu t}\, dt\,. \tag{7}$$

Using integration by parts, the formula (5) and the fact that $xK_0'(x) \to -1$ as $x \to 0^+$ we find that this leads to

$$2\nu z'(\theta) = \frac{4}{\pi} - \frac{8}{\pi} \int_0^\infty e^{2\theta} \sinh 2t K_0(2\, e^\theta \sinh t)\, e^{-2\nu t}\, dt\,. \tag{8}$$

A further integration by parts yields

$$-4\nu^2 z'(\theta) = -\frac{8\nu}{\pi} + \frac{16}{\pi} e^{2\theta} \int_0^\infty \cosh 2t K_0(2\, e^\theta \sinh t)\, e^{-2\nu t}\, dt + \frac{16}{\pi} e^{3\theta} \int_0^\infty \sinh 2t \cosh t K_0'(2\, e^\theta \sinh t)\, e^{-2\nu t}\, dt\,. \tag{9}$$

Now, on the other hand, differentiation of (7) and use of (6) gives

$$z''(\theta) = -\frac{16}{\pi} \int_0^\infty e^{2\theta} \sinh^2 t K_0(2\, e^\theta \sinh t)\, e^{-2\nu t}\, dt$$

and a further differentiation yields

$$z'''(\theta) = -\frac{4}{\pi} \int_0^\infty 8\, e^{2\theta} \sinh^2 t K_0(2\, e^\theta \sinh t)\, e^{-2\nu t}\, dt - \frac{4}{\pi} \int_0^\infty 8\, e^{3\theta} \sinh^3 t K_0'(2\, e^\theta \sinh t)\, e^{-2\nu t}\, dt\,. \tag{10}$$

Now using (10), (7) and (9), respectively, in the first three terms of the left-hand-side shows that

$$z''' + 4\, e^{2\theta} z' - 4\nu^2 z' + 4\, e^{2\theta} z = -8\nu/\pi\,. \tag{11}$$

On the other hand, $J_\nu(e^\theta)$ satisfies

$$y'' + (e^{2\theta} - \nu^2)\, y = 0;$$

hence, using [11, pp. 145–146], we find that

$$u(\theta) = (\mu - \nu)^{-1} J_\mu(e^\theta)\, Y_\nu(e^\theta)$$

satisfies

$$u''' + 2(2\, e^{2\theta} - \mu^2 - \nu^2)\, u' + 4\, e^{2\theta} u = -(\mu + \nu)\, [J_\mu'(e^\theta)\, Y_\nu(e^\theta) - Y_\nu'(e^\theta)\, J_\mu(e^\theta)]\, e^\theta$$

and

$$v(\theta) = (\mu - \nu)^{-1} J_\nu(e^\theta)\, Y_\mu(e^\theta)$$

satisfies

$$v''' + 2(2\, e^{2\theta} - \mu^2 - \nu^2)\, v' + 4\, e^{2\theta} v = -(\mu + \nu)\, [Y_\mu'(e^\theta)\, J_\nu(e^\theta) - J_\nu'(e^\theta)\, Y_\mu(e^\theta)]\, e^\theta\,.$$

Subtracting these two equations letting $\mu \to \nu$ and using [11, p. 76]

$$e^\theta [J_\nu(e^\theta)\, Y_\nu'(e^\theta) - Y_\nu(e^\theta)\, J_\nu'(e^\theta)] = 2/\pi$$

we find that

$$J_\nu(e^\theta)\, \partial Y_\nu(e^\theta)/\partial\nu - Y_\nu(e^\theta)\, \partial J_\nu(e^\theta)/\partial\nu$$

satisfies equation (11). Thus we can assert that

$$J_\nu(e^\theta)\,\partial Y_\nu(e^\theta)/\partial\nu - Y_\nu(e^\theta)\,\partial J_\nu(e^\theta)/\partial\nu =$$

$$= -\frac{4}{\pi}\int_0^\infty K_0(2\,e^\theta \sinh t)\,e^{-2\nu t}\,dt +$$

$$+ A[J_\nu^2(e^\theta) + Y_\nu^2(e^\theta)] + B\{H_\nu^{(1)}(e^\theta)\}^2 + C\{H_\nu^{(2)}(e^\theta)\}^2$$

where the last three terms represent a linear combination of three independent solutions of

$$z''' + 4\,e^{2\theta} z' - 4\nu^2 z' + 4\,e^{2\theta} z = 0\,, \tag{12}$$

the homogeneous equation corresponding to (11). It remains to show that $A = B = C = 0$. From standard asymptotic expansions [11, Chapter 7] we get

$$(\nu - \mu)^{-1}[J_\nu(z)\;Y_\mu(z) - J_\mu(z)\;Y_\nu(z)] =$$

$$= \frac{2}{\pi z}\left\{\frac{\sin(\nu-\mu)\,\pi/2}{\nu-\mu} + O(z^{-1})\right\}, \qquad z \to +\infty\,,$$

and from this it follows that

$$J_\nu(z)\,\partial Y_\nu(z)/\partial\nu - Y_\nu(z)\,\partial J_\nu(z)/\partial\nu = -1/z + O(z^{-2})\,, \quad z \to +\infty\,.$$

At the same time, it follows from [9, p. 339] that

$$-\frac{4}{\pi}\int_0^\infty K_0(2z\sinh t)\,e^{-2\nu t}\,dt = -\frac{1}{z} + O(z^{-2})\,, \qquad z \to +\infty\,.$$

Hence we find, as in [9, p. 341], that

$$\frac{\pi}{2} = \frac{\pi}{2} - A - B\exp\{2i(z - \nu\pi/2 - \pi/4)\} -$$

$$- C\exp\{-2i(z - \nu\pi/2 - \pi/4)\} + o(1),\; z \to \infty.$$

Thus $A = B = C = 0$ and the result is proved in the case where $\nu \neq 0$.

In the case $\nu = 0$, we use the facts (see, e.g., [9, p. 244, Ex. 5.6]) that

$$[\partial J_\nu(z)/\partial\nu]_{\nu=0} = (\pi/2)\;Y_0(z)$$

and

$$[\partial Y_\nu(z)/\partial\nu]_{\nu=0} = -(\pi/2)\;J_0(z)$$

to see that (2) is equivalent to (1).

This research was supported by the National Research Council of Canada.

References

1 Copson, E. T., An Introduction to the Theory of Functions of a Complex Variable, Clarendon Press, Oxford 1935.
2 Elbert, Á., Concavity of the zeros of Bessel functions, Studia Sci. Math. Hungar. **12** (1977), 81–88.
3 Lewis, J. T.; Muldoon, M. E., Monotonicity and convexity properties of zeros of Bessel functions, SIAM J. Math. Analysis **8** (1977), 171–178.
4 Lorch, L., Turánians and Wronskians for the zeros of Bessel functions, SIAM J. Math. Analysis **11** (1980), 223–227.
5 Lorch, L.; Muldoon, M. E.; Szego, P., Some monotonicity properties of Bessel functions, SIAM J. Math. Analysis **4** (1973), 385–392.
6 Lorch, L.; Szego, P., Monotonicity of the differences of zeros of Bessel functions as a function of order, Proc. Amer. Math. Soc. **15** (1964), 91–96.
7 Magnus, W.; Kotin, L., The zeros of the Hankel function as a function of its order, Numer. Math. **2** (1960), 228–244.
8 Naylor, D., On an integral transform, Glasgow Math. J. **20** (1979), 1–14.
9 Olver, F. W. J., Asymptotics and Special Functions, Academic Press, New York and London 1974.
10 Salchev, L. Z.; Popov, V. B., A property of the zeros of cross-product Bessel functions of different orders, ZAMM **56** (1976), 120–121.
11 Watson, G. N., A Treatise on the Theory of Bessel Functions, 2nd ed., Cambridge University Press 1944.
12 Wilkins, J. E., Jr., Nicholson's integral for $J_n^2(z) + Y_n^2(z)$, Bull. Amer. Math. Soc. **54** (1948), 232–234.
13 Willis, D. M., A property of the zeros of a cross-product of Bessel functions, Proc. Cambridge Philos. Soc. **61** (1965), 425–428.

Eingereicht: 20. 1. 1981

Anschrift: Prof. Dr. M. E. Muldoon, Department of Mathematics, York University, Downsview, Ontario M3J 1P3, Canada

ZAMM 61, 599–601 (1981)

R. N. Kaul / B. Gupta

Multi-Objective Programming in Complex Space

In the present paper, the results on linear multiple objective programming due to Isermann ([5]) are presented over polyhedral cones in the complex space. Also the results by the authors ([6]) for the linear fractional vector maximization problem are extended to the complex space.

1. Introduction

The problem of multiple objective linear programming has been studied by a number of authors ([3, 5, 7, 8]) and the concepts of efficiency and proper efficiency have been introduced for the analysis of such problems. On the other hand, linear and nonlinear programming in complex space, also has been discussed by several authors ([1, 2, 4, 9, 10]) and the reference ([1]) contains a number of applications of such an approach. In this note we introduce multiple-objective linear programming in complex space, define the concepts of efficiency and proper efficiency for the same and extend the main result due to Isermann ([5]) over polyhedral cones in the complex space. Moreover, in the last section we further extend this result partially to the linear fractional vector maximization problem by relating the efficient points of it to the efficient points of a linear vector maximization problem in complex space.

2. The problem

We formulate the linear multiple objective programming (MOP) problem in complex space as follows:

Maximize $\mathrm{Re}\,\{c_1^* z, c_2^* z, \ldots, c_p^* z\}$

subject to

$$z \in S = \{z \in \mathbb{C}^n : b - Az \in P,\, z \in Q\}\,,$$

where A is an $m \times n$ matrix having complex entries, the c_i's are constant complex n-vectors, b is a constant complex m-vector and z is a variable complex n-vector; P and Q are polyhedral convex cones in $\mathbb{C}^m$ and $\mathbb{C}^n$, respectively. (The symbol '$*$' denotes the conjugate transpose.) If N is a given convex cone, the polar cone, N^+, of N is defined as

$$N^+ = \{w : z \in N \Rightarrow \mathrm{Re}\,(z, w) \geqq 0\}\,,$$

where the notation (z, w) denotes the usual inner product in complex space. We use the symbol I_p, for a $p \times p$ identity matrix, and e is used for a vector having each entry as unity. The dimension of e shall be clear from the context. Also, let C denote the $p \times n$ criterion matrix, whose rows are formed from the coefficients $c_i^* (1 \leqq i \leqq p)$ of criterion functions that occur in the formulation of (MOP).

3. Definitions

We now give the following definitions for efficiency and proper efficiency.

Definition 1: A point z^0 is said to be an *efficient solution* of (MOP) if $z^0 \in S$ and $\mathrm{Re}\,(c_i^* z) > \mathrm{Re}\,(c_i^* z^0)$ for some $z \in S$ and some index i implies the existence of an index $j \neq i$ such that $\mathrm{Re}\,(c_j^* z) < \mathrm{Re}\,(c_j^* z^0)$.

Definition 2: An efficient solution z^0 of (MOP) is said to be *properly efficient* for (MOP) if there exists a scalar $M > 0$ such that for each $z \in S$ and $\mathrm{Re}\,(c_i^* z) > \mathrm{Re}\,(c_i^* z^0)$ for some index i there is at least one index $j \neq i$, with $\mathrm{Re}\,(c_j^* z) < \mathrm{Re}\,(c_j^* z^0)$ and $\mathrm{Re}\,(c_i^* z - c_i^* z^0) \leqq M\,\mathrm{Re}\,(c_j^* z^0 - c_j^* z)$.

4. Main results

Consider the following linear programming problem (P-I) in complex space

$$\text{Maximize } \mathrm{Re}\left[\begin{pmatrix}0\\e\end{pmatrix}, \begin{pmatrix}z\\w\end{pmatrix}\right]$$

subject to

$$\begin{pmatrix} b \\ -Cz_0 \end{pmatrix} - \begin{pmatrix} A & 0 \\ -C & I_p \end{pmatrix} \cdot \begin{pmatrix} z \\ w \end{pmatrix} \in P \times \{0\}\,, \tag{1}$$

$$\begin{pmatrix} z \\ w \end{pmatrix} \in Q \times L\,, \tag{2}$$

where $z^0 \in S$ and $L = \{w \in \mathbb{C}^p : |\arg w| \leqq \pi/2\}$.

The relation between an efficient solution of (MOP) and the problem (P-I) is outlined in the following lemma.

Lemma 1: *A necessary and sufficient condition for $z^0 \in S$ to be an efficient solution of* (MOP) *is that problem* (P-I) *has an optimal solution* $(\hat{z}, \hat{w})$ *satisfying* $\operatorname{Re} \hat{w}_i = 0$ *for* $i = 1, 2, \ldots, p$.

Proof: Let $z^0 \in S$ be an efficient solution of (MOP). Assume, to the contrary, that $\operatorname{Re} \hat{w}_i \neq 0$ for some $i = k$ (say). Since $\hat{w} \in L$, therefore $\operatorname{Re} \hat{w}_k > 0$ and

$$\operatorname{Re} \hat{w}_i \geqq 0 \quad \text{for} \quad i \neq k\,. \tag{3}$$

Making use of (3) in the constraints of (1) involving the vector w yields the relation

$$\operatorname{Re}(c_i^* \hat{z}) \geqq \operatorname{Re}(c_i^* z^0) \quad \text{for} \quad i \neq k$$

and

$$\operatorname{Re}(c_k^* \hat{z}) > \operatorname{Re}(c_k^* z^0).$$

This contradicts the fact that z^0 is an efficient solution of (MOP).

Conversely, suppose $(\hat{z}, \hat{w})$ is an optimal solution of the problem (P-I) with $\operatorname{Re} \hat{w} = 0$. We assert that z^0 is an efficient solution for (MOP). If not, then there exists a vector $\tilde{z} \in S$, such that

$$\operatorname{Re}(C\tilde{z} - Cz^0) \geqq 0 \quad \text{with} \quad \operatorname{Re}(C\tilde{z}) \neq \operatorname{Re}(Cz^0).$$

This shows that there exists a vector $\tilde{w} = C\tilde{z} - Cz^0$ which belongs to L, such that $(\tilde{z}, \tilde{w})$ is feasible for (P-I) and

$$\operatorname{Re}\left[\begin{pmatrix} 0 \\ e \end{pmatrix}, \begin{pmatrix} \tilde{z} \\ \tilde{w} \end{pmatrix}\right] > 0\,,$$

contradicting the optimality of $(\hat{z}, \hat{w})$. This completes the proof.

The dual of (P-I) is the following problem (D-I):

$$\text{Minimize} \quad \operatorname{Re}\left[\begin{pmatrix} b \\ -Cz^0 \end{pmatrix}, \begin{pmatrix} u \\ v \end{pmatrix}\right]$$

subject to

$$\begin{pmatrix} A^* & -C^* \\ 0 & I_p \end{pmatrix} \begin{pmatrix} u \\ v \end{pmatrix} - \begin{pmatrix} 0 \\ e \end{pmatrix} \in Q^+ \times L^+\,,$$

$$\begin{pmatrix} u \\ v \end{pmatrix} \in P^+ \times \mathbb{C}^p\,.$$

Applying LEVINSON's duality theorem ([9]) as generalized by BEN ISRAEL ([1]), we find that, if $(\hat{z}, \hat{w})$ is a maximizing solution of (P-I), there exists a vector $(\hat{u}, \hat{v})$ which minimizes the dual problem (D-I) such that

$$\operatorname{Re}\left[\begin{pmatrix} b \\ -Cz^0 \end{pmatrix}, \begin{pmatrix} \hat{u} \\ \hat{v} \end{pmatrix}\right] = \operatorname{Re}\left[\begin{pmatrix} 0 \\ e \end{pmatrix}, \begin{pmatrix} \hat{z} \\ \hat{w} \end{pmatrix}\right].$$

An application of Lemma 1 leads to the following result:

Lemma 2: *In order that z^0 be an efficient solution of* (MOP), *it is necessary and sufficient that the problem* (D-I) *has an optimal solution* $(\hat{u}, \hat{v})$ *such that* $\operatorname{Re}[b^*\hat{u} - \hat{v}^*Cz^0] = 0$.

We are now ready to establish the following theorem:

Theorem 1: *In order that z^0 be an efficient solution of* (MOP), *it is necessary and sufficient, that there exists a vector $\xi^0 \in R^p$ such that*

(i) $\xi^0 - e \in L^+$,

(ii) $\operatorname{Re}(\xi^{0*}Cz^0 - \xi^{0*}Cz) \geqq 0$ *for all* $z \in S$.

Proof: Let us assume the existence of a vector $\xi^0 \in R^p$ with the stated properties. Suppose z^0 is not efficient for (MOP). Then there exists a vector $t \in S$, such that

$$w^0 = Ct - Cz^0 \tag{4}$$

satisfies

$$|\arg w^0| \leqq \pi/2 \tag{5}$$

and

$$\operatorname{Re} w_i^0 > 0 \text{ for at least some index } i = k \text{(say)}\,. \tag{6}$$

Thus, we have

$$|\arg(\xi_i^0 - e_i)\, w_i^0| \leqq |\arg(\xi_i^0 - e_i)| + |\arg w_i^0| \leqq \leqq \pi/2 \quad \text{for all} \quad i = 1, 2, \ldots, p\,,$$

where the last inequality follows from the fact that $L^+ = \{z : \arg z = 0\}$.

Hence

$$\operatorname{Re}\{(\xi^0 - e), w^0\} \geqq 0\,. \tag{7}$$

Using (5) and (6), the inequality in (7) reduces to $\operatorname{Re}(\xi^{0*}, w^0) > 0$ i.e.

$$\operatorname{Re}(\xi^{0*}Ct - \xi^{0*}Cz^0) > 0\,.$$

This contradicts the hypothesis (ii).

Conversely, let z^0 be an efficient solution of (MOP). From Lemma 2, there exists a vector $(\hat{u}, \hat{v})$ which optimizes the problem (D-I) and such that

$$\operatorname{Re}(b^*\hat{u}) = \operatorname{Re}(\hat{v}^*Cz^0)\,. \tag{8}$$

It is now easy to establish the optimality of the vector $\hat{u}$, for the following problem (D-2):

Minimize $\operatorname{Re}(b, u)$

subject to

$$A^*u - C^*\hat{v} \in Q^+\,,$$

$$u \in P^+\,,$$

as a consequence of the optimality of $(\hat{u}, \hat{v})$ for (D-I).

The dual to (D-2) is the following linear program (P-2)

Maximize $\operatorname{Re}(C^*\hat{v}, z)$

subject to

$$b - Az \in P\,,$$

$$z \in Q\,.$$

Invoking the duality theorem ([9]) again, we observe that z^0 is optimal for (P-2). This follows, because z^0 is feasible for (P-2) and $\operatorname{Re}(b^*\hat{u}) = \operatorname{Re}(\hat{v}^*Cz^0)$, as established in (8). Hence

$$\operatorname{Re}(\hat{v}, Cz^0) \geqq \operatorname{Re}(\hat{v}, Cz) \quad \text{for all} \quad z \in S,$$

where $\hat{v}$ is the vector satisfying the restriction stated in the theorem.

A consequence of Theorem 1 is the following main result:

Theorem 2: *Let $\xi^0 - e \in L^+$ be such that*

$$\operatorname{Re}(\xi^{0*}Cz^0 - \xi^{0*}Cz) \geqq 0 \quad \textit{for all} \quad z \in S\,.$$

Then z^0 is a properly efficient solution of (MOP).

Proof: It follows from Theorem 1 that z^0 is efficient. We shall show that z^0 is indeed properly efficient for (MOP) with

$$M = (p - 1) \operatorname*{Max}_{i,j} \left(\frac{\xi_j^0}{\xi_i^0}\right).$$

Suppose to the contrary, that for some criterion i and $z \in S$ we have

$$\operatorname{Re}(c_i^*z - c_i^*z^0) > M \operatorname{Re}(c_j^*z^0 - c_j^*z) \quad \text{for all} \quad j \neq i$$

such that

$$\operatorname{Re}(c_j^*z^0) > \operatorname{Re}(c_j^*z)\,.$$

Therefore, by taking $M = (p - 1) \operatorname*{Max}_{i,j} \left(\frac{\xi_j^0}{\xi_i^0}\right)$, it follows that

$$\operatorname{Re}(c_i^*z - c_i^*z^0) > (p - 1) \frac{\xi_j^0}{\xi_i^0} \operatorname{Re}(c_j^*z^0 - c_j^*z) \quad \text{for all} \quad j \neq i\,.$$

Multiplying throughout by $\dfrac{\xi_i^0}{(p - 1)}$ and summing over all indices $j \neq i$, we get

$$\operatorname{Re}\{\xi_i^0(c_i^*z - c_i^*z^0)\} > \operatorname{Re}\Big\{\sum_{j \neq i} \xi_j^0(c_j^*z^0 - c_j^*z)\Big\}$$

i.e.,

$$\operatorname{Re}[\xi^{0*}Cz - \xi^{0*}Cz^0] > 0\,,$$

a contradiction to our assumption. Hence z^0 is properly efficient solution of (MOP).

Thus, we have generalized the main result of ISERMANN ([5]) to the complex space, namely that an efficient solution is properly efficient for the linear multiple objective program in the complex space as well.

5. Extensions

We now partially extend the result established in Section 4 to the linear fractional vector maximization problem in the complex case. We obtain a corresponding condition as in the linear case for a point to be efficient, which in this case, however, is only a necessary condition, not necessarily sufficient. This necessary condition has already been established for the linear fractional problem in the real space by the authors ([6]).

This linear fractional vector maximization (LFMP) problem in complex space is:

Maximize $h(x)$

where

$$h(x) = \left\{\frac{\operatorname{Re} c_1^* z}{\operatorname{Re} d_1^* z}, \frac{\operatorname{Re} c_2^* z}{\operatorname{Re} d_2^* z}, \dots, \frac{\operatorname{Re} c_p^* z}{\operatorname{Re} d_p^* z}\right\}$$

subject to

$$z \in S = \{z \in C^n : b - Az \in P, z \in Q\},$$

where the different symbols used have the same meaning as described earlier. In addition, the d_i's are constant complex n-vectors for all $i = 1, 2, \dots, p$. It is also assumed that

$$\operatorname{Re} d_i^* z > 0 \quad \text{for all} \quad i \in \{1, 2, \dots, p\} \quad \text{and all} \quad z \in S .$$

Here, the definition of efficiency becomes

Definition 3: A point z^0 is said to be *efficient for* (LFMP), if $z^0 \in S$ and

$$\frac{\operatorname{Re} c_i^* z}{\operatorname{Re} d_i^* z} > \frac{\operatorname{Re} c_i^* z^0}{\operatorname{Re} d_i^* z^0}$$

for some index i and $z \in S$ implies the existence of an index $j \neq i$ such that

$$\frac{\operatorname{Re} c_j^* z}{\operatorname{Re} d_j^* z} < \frac{\operatorname{Re} c_j^* z^0}{\operatorname{Re} d_j^* z^0} \quad \text{for all} \quad z \in S .$$

To the above (LFMP) in complex space we associate the following linear vector maximization (LVMP) problem, in complex space, namely,

$$\text{Maximize} \quad \operatorname{Re} \{c_1^* z, c_2^* z, \dots, c_p^* z, -d_1^* z, \dots, -d_p^* z\}$$

subject to

$$z \in S .$$

The following theorem gives a relation between (LFMP) and (LVMP).

Theorem 3: *If $z^0 \in S$ is an efficient solution of problem* (LFMP), *then it is efficient for* (LVMP).

Proof. Let z^0 be efficient for (LFMP). Further, suppose to the contrary that z^0 is not efficient for (LVMP). Then there would exist a $z \in S$ and a certain index $j \in \{1, 2, \dots, p\}$, such that

$$[\operatorname{Re} c_i^* z \geqq \operatorname{Re} c_i^* z^0 \quad \text{and} \quad \operatorname{Re} d_i^* z \leqq \operatorname{Re} d_i^* z^0 \quad \text{for all } i]$$

and

$$[\operatorname{Re} c_j^* z > \operatorname{Re} c_j^* z^0 \quad \text{or} \quad \operatorname{Re} d_j^* z < \operatorname{Re} d_j^* z^0],$$

which implies that

$$\frac{\operatorname{Re} c_i^* z}{\operatorname{Re} d_i^* z} \geqq \frac{\operatorname{Re} c_i^* z^0}{\operatorname{Re} d_i^* z^0} \quad \text{for all } i$$

and

$$\frac{\operatorname{Re} c_j^* z}{\operatorname{Re} d_j^* z} > \frac{\operatorname{Re} c_j^* z^0}{\operatorname{Re} d_j^* z^0},$$

a contradiction to the assumption that z^0 is efficient for (LFMP). Hence the theorem.

We now establish the necessary condition in the following theorem.

Theorem 4: *If z^0 is efficient for* (LFMP), *then it is necessary that there exist $\xi^0 \in R^p$ and $\eta^0 \in R^p$ with $\xi^0 - e \in L^+$ and $\eta^0 - e \in L^+$, such that*

$$\operatorname{Re} (\xi^{0*} C z^0 - \eta^{0*} D z^0) \geqq \operatorname{Re} (\xi^{0*} C z - \eta^{0*} D z) \quad \textit{for all} \quad z \in S ,$$

where D is the criterion matrix whose rows are formed from the coefficients d_i^* $(1 \leqq i \leqq p)$ which occur in the formulation of (LFMP).

Proof: As z^0 is efficient for (LFMP), theorem 3 shows that it is efficient for (LVMP), too. Making use of Theorem 1, we find that there exists a vector $(\xi^0, \eta^0) \in R^{2p}$, such that

$$\xi^0 - e \in L^+ , \qquad \eta_0 - e \in L^+ ,$$

and

$$\operatorname{Re} (\xi^{0*} C z^0 - \eta^{0*} D z^0) \geqq \operatorname{Re} (\xi^{0*} C z - \eta^{0*} D z) \quad \text{for all} \quad z \in S .$$

This completes the proof.

Acknowledgement

The authors wish to express their deep gratitude to the referee for his helpful suggestions and comments.

References

1 BEN ISRAEL, A., Linear equations and inequalities on finite dimensional, real or complex, vector spaces: A unified theory, J. Math. Analysis Appl. **27** (2), 367–389 (1969).
2 BHATIA, D.; KAUL, R. N., Non-linear programming in complex space, J. Math. Analysis Appl. **28**, 144–152 (1969).
3 GEOFFRION, A. M., Proper efficiency and the theory of vector maximization, J. Math. Analysis Appl. **22**, 618–630 (1968).
4 HANSON, M. A.; MOND, B., Quadratic programming in complex space, J. Math. Analysis Appl. **20** (1967).
5 ISERMANN, H., Proper efficiency and the linear vector maximum problem, Operations Res. **22** (1), 189–191 (1974).
6 KAUL, R. N.; BINA GUPTA, Efficiency and linear fractional vector maximum value problem, ZAMM **60**, 112–113 (1980).
7 KLINGER, A., Improper solutions of the vector maximization problem, Operations Res. **15**, 570–572 (1967).
8 KUHN, H. W.; TUCKER, A. W., Non-linear programming, in J. NEYMANN (ed), Proc. Second Berkeley Symposium on Math. Statistics and Probability, University of California (1951), pp. 481–492.
9 LEVINSON, N., Linear programming in complex space, J. Math. Analysis Appl. **14**, 44–62 (1966).
10 MOND, B., Duality for a complex non-linear program, Opsearch **11** (1), 1–9 (1974).

Eingereicht: 11. 7. 1980, revidierte Fassung: 9. 2. 1981

Anschrift: Mrs. BINA GUPTA, Professor Dr. R. N. KAUL, Department of Mathematics, University of Delhi, Delhi-110007, INDIA

BUCHBESPRECHUNGEN

Kallianpur, G., Stochastic Filtering Theory, Applications of Mathematics 13, Berlin-Heidelberg-New York, Springer-Verlag 1980, XVI, 316 S., DM 59,50, US $ 35.10. ISBN 3-540-90445-X

Die stochastische Filtertheorie findet sowohl im Rahmen der theoretischen Forschungen in der Stochastik als auch in den verschiedensten praktischen Anwendungsgebieten, insbesondere der Naturwissenschaften, eine zunehmende Beachtung. Vorliegende Arbeit widmet sich vor allem einer auf mathematisch hohem Niveau stehenden Darlegung theoretischer Ergebnisse der genannten Problematik verbunden mit einer ausführlichen Betrachtung von in diesem Zusammenhang interessierenden Klassen stochastischer Prozesse. Ausgehend von einigen Grundlagen über Maßtheorie und stochastische Prozesse werden der Wienersche Prozeß (u. a. Reihendarstellungen, Trajektorieneigenschaften) und Martingale (u. a Zerlegungssatz von Supermartingalen, quadratisch integrierbare Martingale, lokale Martingale) behandelt. Mit der Erläuterung der stochastischen Integrale (Integrale bezüglich stetigen quadratisch integrierbaren Martingalen und stetigen lokalen Martingalen speziell auch des Ito-Integrals) sowie der Behandlung der Ito-Formel für den allgemeinen Fall stetiger lokaler Halbmartingale (sowohl eindimensional als auch mehrdimensional) werden wichtige Resultate bereitgestellt, die zum Verständnis des eigentlichen Anliegens der Arbeit notwendig sind. Daran anschließend geht der Verfasser auf die wichtigsten Aspekte der sogenannten Ito-Theorie ein, und zwar auf stochastische Ito-Differentialgleichungen (u. a. Existenz, Eindeutigkeit und Eigenschaften der Lösung, lineare stochastische Differentialgleichungen, Momente), Funktionale des Wienerschen Prozesses (multiples Wiener-Integral und Anwendungen) und die Absolutstetigkeit von Maßen und Radon-Nikodym-Ableitungen (Ergebnisse von Cameron, Martin und Girsanov zur Transformation Wienerscher Prozesse). Die hier diskutierten Resultate sind insbesondere in der nichtlinearen Filtertheorie praktisch bedeutsam. Diese wird vom Verfasser sehr ausführlich betrachtet lineare Filter werden als Spezialfall der nichtlinearen verhältnismäßig kurz abgehandelt. Wesentliches Resultat ist die Herleitung einer stochastischen Gleichung für das allgemeine nichtlineare Filterproblem, die anhand von Beispielen anschaulich erläutert wird. Von praktischem Interesse auch für lineare Filter dürfte das Kapitel über Gaußsche Lösungen von stochastischen Gleichungen sein, obwohl es durch seine operatoren-theoretischen Betrachtungen etwas aus dem Rahmen der durchgängig verwendeten Martingalmethoden fällt. Selbstverständlich beinhaltet die Arbeit auch eine abgerundete Darstellung zur Kalman-Bucy-Filtertheorie, die als Spezialfall der nichtlinearen Theorie herausgearbeitet wird. Insgesamt kann eingeschätzt werden, daß sich vorliegende Arbeit insbesondere an Studenten der Wahrscheinlichkeitstheorie und an Naturwissenschaftler wendet, deren Forschungsgebiet Fragen der Filtertheorie beinhaltet. Zum Verständnis der Ausführungen sind allerdings Grundkenntnisse aus der Maßtheorie, der Funktionalanalysis und der Wahrscheinlichkeitstheorie unbedingt erforderlich. Eine klare Gliederung, übersichtliche Gestaltung, sorgfältige Auswahl der Bezeichnungen sowie eine exakte Darstellungsweise entsprechen dem hohen Niveau der vorgestellten theoretischen Ergebnisse.

Berlin C. Lange

Winfree, A. T., The Geometry of Biological Time. Biomathematics 8. Berlin-Heidelberg-New York, Springer-Verlag 1980. XIV, 530 S., 290 Abb., DM 59,50. US $ 32.80. ISBN 3-540-09373-7

Eine Vielzahl von Experimenten zeigt, daß die meisten periodisch ablaufenden biologischen Vorgänge ihre Periodizität auch dann mit relativ stabiler Frequenz beinhalten, wenn die äußeren periodischen Einflüsse (z. B. Wechsel von Tag und Nacht, Jahreszeiten) wegfallen. Die Natur der „inneren biologischen Uhren" wird intensiv erforscht, ist aber besonders im Bereich längerer Perioden (Tage und darüber) noch weitgehend ungeklärt.

Im vorliegenden Buch werden Experimente mit biologischen Uhren beschrieben. Hauptgegenstand der Untersuchungen sind Populationen gleichartiger Oszillatoren mit Phasenverschiebungen. Verschiedene Auftragungen der Phasen gegen die phasenverschiebenden Stimuli führen zu interessanten Flächen im R^3 (größtenteils mit Singularitäten). Der Autor versucht nicht die biologischen Uhren durch mathematische Gleichungen zu modellieren. Seine „Experimente mit Uhren und Abbildungen" sollen dem Leser einen Eindruck von der Vielfalt und Schönheit der Problematik und Anregungen zur weiteren Forschung vermitteln. Er bietet keine fertigen theoretischen Erklärungen an. Das Buch ist nicht vorrangig für Mathematiker geschrieben. Die verwendeten geometrischen Hilfsmittel werden ausführlich erläutert.

Das Buch ist nicht nur für Spezialisten der angesprochenen Thematik sehr informativ, es ist wegen der anschaulichen Darstellung wichtiger experimenteller Befunde auch dem naturwissenschaftlich allgemein interessierten Leser zu empfehlen.

Berlin W. Meiske

Amazigo, J. C. / Rubenfeld, L. A., Advanced Calculus and Its Applications to the Engineering and Physical Sciences. New York-Chichester-Brisbane-Toronto, John Wiley and Sons 1980. VIII, 407 S., £ 10.25. ISBN 0-471-04934-4

This book is an excellent introduction to advanced calculus for those being interested in engineering sciences and physics as well as applied mathematics. Readers should be familiar with elementary calculus, linear algebra and (ordinary) differential equations. On this basis the authors present the calculus for functions of several variables, i. e. continuity, curves, surfaces, parametric representations, the differential calculus, extrema without and with constraints, integrals of the different kinds used in physics (multiple, surface, line integrals, the famous theorems in this field etc.), vectors and vector fields, i. e. limits, continuity, differential operations and the corresponding integral theorems, calculus of variations as a natural continuation of calculus, infinite series of functions, i. e. different kinds of convergence, power and Taylor series and applications to differential equations and finally a chapter on partial differential equations. — The book results from courses on advanced calculus held by the authors during more than 20 years and certifies their great experiences. There is always an equilibrium between mathematical rigor and intuition how to apply mathematics in a successful way, thereby being neither a potpourri of results or even receipts nor pure mathematics with some physical examples as a sign-board for applications. All sections contain examples drawn from physics, theoretical mechanics or engineering. The knowledge on the subject treated in a section can be checked and deepened by the reader himself by means of numerous problems and exercises the results of which are collected at the end of the book, where the reader in addition can find some examples of orthogonal coordinate systems, certain vector relations, some differential equations of mathematical physics and a review of ordinary differential equations. To conclude with the book can highly be recommended to graduate students and researchers as well being interested to become familiar with results and methods of advanced calculus and to apply them to physical or engineering problems.

Berlin H. Weinert

Boiti, M. / Pempinelli, F. / Soliani, G. (eds.), Nonlinear Evolution Equations and Dynamical Systems, Proceedings, Lecce 1979, Lecture Notes in Physics 120, Berlin-Heidelberg-New York, Springer-Verlag 1980, VI, 368 S., 25 Abb., DM 43,–, US $ 25.40. ISBN 3-540-09971-9

In diesem Band sind die Vorträge einer von der Universität Lecce (Italien) veranstalteten Tagung über die Bedeutung nichtlinearer Evolutionsgleichungen für die Theorie dynamischer Systeme (und umgekehrt) zusammengestellt. Einmal sind sie zu sehen als in engem Kreis gehaltene Lektionen über die bekannten Zusammenhänge zwischen der inversen Streutransformation

(Spektraltransformationsmthode), Solitonen, Bäcklundtransformationen und Erhaltungsgrößen bei vollständig integrablen Evolutionsgleichungen. Einige Beiträge stellen eine nützliche Einführung in die Standardliteratur zu diesen Problemen dar. Das Erscheinen in den „Lecture Notes in Physics" mit dem Anliegen, neue Fortschritte und Trends anzugeben und eine repräsentative Auswahl der letzten Arbeiten auf diesem Gebiet zu sammeln, wird jedoch nur begrenzt erfüllt. Hiervon kann man sich sehr leicht überzeugen durch einen Vergleich mit dem Bd. 23 „Solitons" (1980) der „Topics in Current Physics" (R. K. BULLOUGH und P. J. CAUDREY (Eds.)). Dieser Band setzt sich mit den schwierigen Problemen, das oben genannte Ziel für die Solitonentheorie zu erreichen, weit kritischer auseinander und ist dadurch vielleicht diesem Ziel schon sehr nahe gekommen.

Hervorzuheben sind einige Beiträge, unter ihnen der von SASAKI und BULLOUGH zur geometrischen Interpretation der Methode der inversen Streutransformation, vor allem im Hinblick auf die weiterführenden modernen differentialgeometrischen Formulierungen (vgl. Lecture Notes in Mathematics 810 (1980), R. MARTINI, Ed.). CHUDNOVSKY und CHUDNOVSKY behandeln Operatorformulierungen vollständig integrierbarer Evolutionsgleichungen, abgeleitet durch Verallgemeinerung des Riemann-Hilbert-Problems. Ähnlich wie bei ZAKHAROV in dem oben zitierten Band 23 „Solitons" gelingen dadurch mehrdimensionale Erweiterungen solcher Evolutionsgleichungen. JACOB und STERNBERG gehen von den „Poisson-Strukturen" aus, die mit nichtlinearen Evolutionsgleichungen verknüpft sind und finden über eine Algebra formaler Pseudodifferentialoperatoren einen Zugang zur Bedingung der vollständigen Integrabilität. PIRANI gibt eine knappe Darstellung der Bäcklund-Transformationen im Rahmen der Theorie der Abbildungen von Faserbündeln, TER HAAR beschreibt kurz die eigentlich doch nicht recht wirkungsvolle Rolle des Solitonenkonzepts in der Theorie der starken Langmuir-Turbulenz.

Weitere Beiträge sind anderen Aspekten der Bäcklund-Transformationen, der IST und der vollständigen Integrabilität von Evolutionsgleichungen, den Erhaltungsgrößen, den Zusammenhängen zwischen Ähnlichkeitslösungen, und konformen Abbildungen bei Wirbelbewegungen sowie Anwendungen auf Josephson-Tunnel-Kontakte gewidmet.

Berlin G. WALLIS

Landau, L. D. / Lifschitz, E. M., Lehrbuch der Theoretischen Physik IV, Relativistische Quantentheorie (Berestetzki, W. B. / Lifschitz, E. M./Pitajewski, L. P.), 4., bearb. Auflage, Berlin, Akademie-Verlag 1980, XII, 586 S., 24 Abb., DM 38,–. BN 7625470 (5436/4)

Der vorliegende Band IV der bekannten Lehrbuchreihe von L. D. LANDAU und E. M. LIFSCHITZ über theoretische Physik behandelt Fragen der relativistischen Quantenfeldtheorie. Es ist die vierte deutsche Auflage, die die bisher getrennt erschienenen Teilbände IVa und IVb vereinigt. In 14 Kapiteln werden in der Hauptsache Probleme der Quantenelektrodynamik behandelt, wobei den physikalischen Fragestellungen der Vorrang gegeben wird und die Behandlung konkreter Anwendungen einen breiten Raum einnimmt. Nach Quantisierung der freien Felder werden die Bewegung eines Teilchens im äußeren Feld und einfache Strahlungsprobleme behandelt, darunter Zeeman- und Stark-Effekt, das Wasserstoffatom und der Photoeffekt. Im Rahmen der kovarianten Störungstheorie werden Wechselwirkungsprozesse zwischen Elektronen und Photonen untersucht, so die Compton-Streuung, Paarerzeugung und -vernichtung von Elektronen-Positronen und die Bremsstrahlung. Ein Kapitel ist neben Renormierungsproblemen der Diskussion einiger allgemeiner, von der Störungstheorie unabhängiger Eigenschaften von Greenschen Funktionen und Streuamplituden gewidmet. Sehr eingehend werden Strahlungskorrekturen behandelt, so das anomale magnetische Moment des Elektrons und der Lamb-Shift, wo die Quantenelektrodynamik ihre größten Erfolge aufzuweisen hat. In der Tat gibt es bis heute keine Theorie, die auch nur annähernd eine solche hohe Genauigkeit in der Übereinstimmung mit experimentellen Daten aufzuweisen hat. Abschließend werden in zwei Kapiteln asymptotische Formeln für die Quantenelektrodynamik und die Elektrodynamik der Hadronen behandelt.

Die Darstellung des Stoffes ist klar und ausgewogen, doch wird sie dem gegenwärtigen Stand der relativistischen Physik nicht gerecht. Die Forschung der letzten 15 Jahre hat hier zu ganz wesentlichen Weiterentwicklungen geführt, wie zur einheitlichen Theorie der elektromagnetischen und schwachen Wechselwirkung von WEINBERG und SALAM (Nobelpreis 1979) und zur sogenannten Quantenchromodynamik als Theorie der starken Wechselwirkung (von der die Autoren vermuten, daß dazu prinzipiell neue physikalische Vorstellungen nötig wären). Letztere Theorie stellt eine bestimmte Verallgemeinerung der Quantenelektrodynamik dar und beschreibt die Wechselwirkung der Quarks und Gluonen, der fundamentalen Teilchen der starken Wechselwirkung. Diese modernen Entwicklungen haben zu einer einheitlichen Auffassung für alle Wechselwirkungen geführt, nämlich zur Erkenntnis, daß diese allesamt durch Eichfeldtheorien beschrieben werden. Eine Neuauflage dieses Bandes wird dem unbedingt Rechnung tragen müssen, wenn sie dem (sehr anspruchsvollen) Gegenstand „Relativistische Quantentheorie" gerecht werden soll.

Berlin F. KASCHLUHN

Lamb, G. L., jr., Elements of Soliton Theory, New York-Chichester-Brisbane-Toronto, John Wiley & Sons 1980, XII, 289 S., £ 16.00. ISBN 0-471-04559-4

Mit diesen „Elementen der Solitonentheorie" von LAMB ist nach einer Reihe Übersichtsartikeln und Tagungsbänden, nach der „Theorie der Solitonen" (in Russ.) von ZAKHAROV, MANAKOV und PITAYEVSKA (1980) sowie dem Bd. 23 „Solitons" (1980) von „Topics in Current Physics" (BULLOUGH und CAUDREY, Eds.) nun auch die erste echte Monographie zu dieser Thematik erschienen. Die Beschränkung auf die Elemente der Solitonentheorie drückt sich einerseits darin aus, daß auf die Neu- oder Umformulierungen mit Hilfe der modernen differentialgeometrischen und gruppentheoretischen Methoden (vergl. R. MARTINI „Geometrical Approaches to Differential Equations" (1980), Lecture Notes in Mathematics, Bd. 810) verzichtet wird, wie vom Autor einleitend betont. Ein zweiter wesentlicher Zug in dieser Richtung scheint die bewußte Anschließung der nichtlinearen Evolutionsgleichungen an ein zugehöriges lineares Sturm-Liouville-Problem. So soll ein möglichst einfacher und direkter Zusammenhang zwischen nichtlinearen partiellen Differentialgleichungen (DG) mit Solitonenlösungen und linearen gewöhnlichen DG hergestellt werden.

Multisolitonenlösungen werden als Bargmann-Potentiale eingeführt. Die am Schluß abgehandelte und zu einem wesentlichen Teil auf Originalarbeiten des Autors beruhende Verknüpfung der Solitonengleichungen mit Bewegungsgleichungen von Raumkurven (Filamenten) versucht, die ad hoc eingeführten und den Solitonengleichungen zugeordneten (linearen) Eigenwertprobleme in die geometrische Interpretation mit einzubeziehen. Diese Betrachtung von „Solitonen auf einem Wirbelfaden" gelingt jedoch nur für einige der Solitonengleichungen (Nichtlineare Schrödingergleichung, modifizierte Korteweg-de Vries-Gleichung, sinus-Gordon-Gleichung).

Die Abhandlung der Methode der Inversen Streutransformation wird in Standardform durchgeführt. Bei den Anwendungen werden, ebenfalls entsprechend dem Arbeitsgebiet des Autors, die Solitonen in der nichtlinearen Optik (Ausbreitung von kurzen Impulsen bei der Wechselwirkung intensiver Lichtstrahlung mit Medien verschiedener Aggregatzustände) sehr gründlich hinsichtlich ihrer theoretischen Grundlagen, aber auch sehr praxisnahe dargestellt. Auf diese Weise wird ein interessanter Einblick in die Arbeit der „Arizona-Gruppe" der Solitonenforschung verknüpft mit einer geschickten Einführung in die grundlegenden Zusammenhänge der klassischen Solitonenphänomene und mit einigen typischen Beispielen ihrer physikalischen Anwendung.

Berlin G. WALLIS

Collatz, L. / Meinardus, G. / Werner, H. (Eds.), Numerical Methods of Approximation Theory. Vol. 5. Vortragsauszüge der Tagung in Oberwolfach, März 1979. ISNM 52. Basel-Boston-Stuttgart, Birkhäuser Verlag 1980. 337 S., sFr. 45,–. ISBN 3-7643-1103-7

Die Vortragsauszüge der Oberwolfacher Tagungen über Numerische Methoden der Approximationstheorie, die regelmäßig alle zwei Jahre stattfinden, geben einen guten Überblick über Entwicklungstendenzen in der Approximationstheorie. Dieser Überblick kann naturgemäß nicht vollständig sein, aber er ermöglicht eine schnelle Orientierung über aktuelle Forschungsgebiete und neue Forschungsergebnisse. Die Beschäftigung mit Problemen der Approximationstheorie ist unverändert reizvoll, da dabei dem Mathematiker sowohl auf dem Gebiet der Theorie als auch auf dem der Anwendungen einschließlich der Rechentechnik ein reiches Betätigungsfeld geboten wird. Aus diesem Grunde überdecken die vorliegenden 22 Beiträge ein weites Spektrum, in dem sich aber einige Schwerpunkte erkennen lassen. Im folgenden wird versucht, die einzelnen Beiträge diesen Schwerpunkten zuzuordnen.

Mehrdimensionale Approximationen: DAHMEN, W., Konstruktion mehrdimensionaler B-Splines und ihre Anwendung auf Approximationsprobleme; GOLITSCHEK v., M., Approximation of functions of two variables by the sum of two functions of one variable; HANDSCOMB, D. C., Some problems in the generation of smooth surfaces; MEHAUTE LE, A., Prolongement d'un champ taylorien connu sur les cotes d'un triangle; SCHERER, R. und ZELLER, K., Gestuft Approximation in zwei Variablen.

Eindimensionale Apropximationen, Splines und spezielle Funktionen: BRAESS, D., Über die numerische Behandlung von Prolemen mit kleinen Nennern; BRASS, H., Approximation durch Teilsummen von Orthogonalpolynomen; FERGUSON, L. B. D., Approximation by polynomials with integral coefficients and digital filter design; McLEAN, R. A. and WATSON, G. A., Numerical methods for nonlinear discrete L_1 approximation problems; MEINARDUS, G. und MERZ, G., Hermite-Interpolation mit periodischen Spline-Funktionen; NEUMANN, E., Convex interpolating splines of arbitrary degree; PEDERSEN, P. W., Some approximations for trigonometrical functions; SCHABACK, R., Bemerkungen zur Fehlerabschätzung bei linearer Tschebyscheff-Approximation; SCHEMPP, W., Approximation und Transformationsmethoden II; SCHMEISSER, G., Two contributions to numerical quadrature; WERNER, H., Ein Algorithmus zur rationalen Interpolation.

Optimierung und Approximation: BLATT, H.-P., Lipschitz-Stabilität von Optimierungs- und Approximationsaufgaben; KOSCIELNY, E. und KRABS, W., Approximation problems in control theory.

Die numerischen Akzente der Tagung werden unterstrichen durch Beiträge zur Lösung nichtlinearer Gleichungssysteme (HACKL, H.; WACKER, H. J.; ZULEHNER, W., Aufwandsoptimale Schrittweitensteuerung bei Einbettungsmethoden) und zur Lösung von Anfangswertaufgaben bei gewöhnlichen Differentialgleichungen (BÖHMER, K. und FLEISCHMANN, H.-J., Self-adaptive discrete Newton methods for Runge-Kutta methods), während eine Arbeit zur Bernsteinschen Vermutung über die Charakterisierung optimaler Operatoren (LOEB, H. L., A differential equation approach to the Bernstein problem) ausgesprochen klassischen Fragestellungen zuzuordenen ist.

Selbstverständlich ist der Teilnehmerkreis einer internationalen Tagung auch gewissen Zufälligkeiten unterworfen, aber die bewährten Leiter der Tagung, die gleichzeitig Herausgeber dieser Berichte sind, werden sicher in Zukunft für eine glückliche Kombination von Theorie und Anwendung der Approximationstheorie Sorge tragen.

Dresden H. MÜHLIG

Kovács, L. B., Combinatorial Methods of Discrete Programming, Mathematical Methods of Operations Research 2, Budapest, Akadémiai Kiadó 1980, 283 S., ISBN 963 05 2004 4

Das vorliegende Buch ist sowohl ein Lehrbuch als auch eine Monographie der diskreten Optimierung, die aus Lecture Notes aus dem Jahre 1969 entstanden ist. Im ersten Kapitel werden häufig auftretende Optimierungsmodelle vorgestellt. Sodann werden implizite Enumerationsalgorithmen sowie Branch und Bound Verfahren ausführlich behandelt. Im vierten Kapitel wird die dynamische Optimierung als Lösungsmöglichkeit für Rucksackprobleme und diskrete lineare Programme vorgestellt. In diesem Zusammenhang wird auch auf Gruppenprobleme eingegangen. Anschließend wurden surrogate constraints sowie Benders' Dekomposition besprochen, wie auch die Balasschen Enumerationsalgorithmen neu dargestellt. Ein wichtiger Abschnitt behandelt Heuristiken in der Diskreten Optimierung obwohl gerade hier in den letzten Jahren eine Fülle von Ergebnissen gewonnen wurde, die unerwähnt bleiben. Ferner wird eine neue Technik zur Lösung des Set Covering Problems vorgestellt und der Autor berichtet über seine Erfahrungen mit komplexen Algorithmen. Ausführlich dargestellte Beispiele illustrieren die besprochenen Verfahren.

Im letzten Abschnitt werden stichwortartig ausgewählte neuere Arbeiten auf dem Gebiete der diskreten Optimierung charakterisiert. Dabei bleiben jedoch einige Gebiete, die in den letzten Jahren besonderes Interesse in der Forschung fanden wie Komplexitätstheorie, effiziente Algorithmen für spezielle Probleme, probabilistische Analysis von Algorithmen unerwähnt. Eine Bibliographie über neuere Arbeiten beschließt das Buch.

Köln R. E. BURKARD

In der DDR beziehbar im Rahmen des Kontingents.

EINGEGANGENE BÜCHER

Bei der Schriftleitung sind folgende Bücher eingegangen (ausführliche Besprechung bleibt vorbehalten).

Rudin, W., Analysis. Weinheim, Physik Verlag 1980. XI. 335 S., DM 48,–. ISBN 3-87664-039-3 (Übers. aus dem Englischen)

Reid, W. T., Sturmian Theory for Ordinary Differential Equations. Applied Mathematical Sciences 31. Berlin-Heidelberg-New York, Springer-Verlag 1980. XV, 559 S., DM 54,–. US $ 31.90. ISBN 3-540-90542-1

Landau, L. D. / Lifschitz, E. M., Lehrbuch der Theoretischen Physik VIII: Elektrodynamik der Kontinua. 3. Auflage. Berlin, Akademie-Verlag 1980. X, 480 S., M 32,–. BN 760-245-5 (5436/VIII)

Doolan, E. P. / Miller, J. J. H. / Schilders, W. H. A., Uniform Numerical Methods for Problems with Initial and Boundary Layers. Dublin, Boole Press 1980. XII, 425 S., £ 28.00. US $ 60.00. ISBN 0-906783-01-1

Mees, A. I., Dynamics of Feedback Systems. Chichester-New York-Brisbane-Toronto, John Wiley & Sons 1981. X, 214 S., £12.95. ISBN 0-471-27822-X

Ben-Israel, A. / Ben-Tal, A. / Zlobec, S., Optimality in Nonlinear Programming. A Feasible Directions Approach. New York-Chichester-Brisbane-Toronto, John Wiley & Sons 1981. XII, 114 S., £ 12.50. ISBN 0-471-08057-8

Topics in Modern Operator Theory. 5th International Conference on Operator Theory, Timişoara and Herculane (Romania), June, 1980. Eds.: Apostol, C./Douglas, R. G./Sz.-Nagy, B./Voiculescu, D.. Basel-Boston-Stuttgart, Birkhäuser Verlag 1981. 336 S., sFr. 42.–/DM 46,–/US $ 28.00. ISBN 3-7643-1244-0

Jentsch, L. / Maul, J., Zur Elastizitäts- und Thermoelastizitätstheorie. Mathematische Forschung 4. Berlin Akademie-Verlag 1980. 157 S., M 20,–. BN 762 8102 (2182/4)

FSC
www.fsc.org
MIX
Papier | Fördert gute Waldnutzung
FSC® C083411

Zeitfracht Medien GmbH
Ferdinand-Jühlke-Straße 7
99095 Erfurt, Deutschland
produktsicherheit@kolibri360.de